Robot Technology

Volume 1: Modelling and Control

Volume 2: Interaction with the Environment

Volume 3: Teleoperations

Volume 4: Robot Components

Volume 5: Robotic Languages and Programming Methods

Volume 6: Decision Autonomy and Artificial Intelligence

Volume 7: Performance and Computer-aided Design

Robot Technology

Series Editors and Consultants: Philippe Coiffet and Pierre-Jean Richard
English Language Series Consultant:
Professor I. Aleksander, Brunel University, Uxbridge, England

Volume 2

INTERACTION WITH THE ENVIRONMENT

Philippe Coiffet
Automation Laboratory, Montpellier, France

Kogan Page
London

Prentice-Hall, Inc.
Englewood Cliffs, NJ 07632

Library of Congress Catalog Card Number 83-3091

Production Supervision: Karen Skrable
Manufacturing Buyer: Gordon Osbourne
Cover Design: Wanda Lubelska
Cover Photo: Courtesy of Cincinnati Milacron

Printed in the United States of America

10 9 8 7 6 5 4 3

ISBN 0-13-782128-X

Prentice-Hall International, Inc., *London*
Prentice-Hall of Australia Pty. Limited, *Sydney*
Editora Prentice-Hall do Brasil, Ltda., *Rio de Janeiro*
Prentice-Hall Canada Inc., *Toronto*
Prentice-Hall of India Private Limited, *New Delhi*
Prentice-Hall of Japan, Inc., *Tokyo*
Prentice-Hall of Southeast Asia Pte. Ltd., *Singapore*
Whitehall Books Limited, *Wellington, New Zealand*

Contents

Introduction

Robots are in the news. In the development of new technology they are seen as a target for advancing techniques, whilst the media are relentlessly persuading the layman that this is 'the age of the robot'. But it is only recently that the subject has begun to emerge as a coherent but multi-disciplinary activity, the result of a coming-together of work on different aspects of the subject.

The first phase of development of robots as a self-sufficient discipline could be labelled *programmable automation*, and was the subject of the first volume in this series.

Another concept often advanced by the media is that, through the capability offered by 'fifth generation' computing, we shall approach the next century equipped with the *intelligent* or *advanced* robot. This has given rise to much debate, particularly regarding the application of artificial intelligence (AI) techniques. In this volume the concentration is on important advances that do not rely on AI but are instrumental in the creation of the advanced robot. Such a device is capable of taking relatively sophisticated measurements of its surroundings and following them with simple, but effective, action.

Development and application of the robot

Despite the fact that some industrialized countries have managed to develop highly productive manufacturing systems without the help of robots, there are specific areas of application that would clearly benefit from their rapid development. The main ones are:

1. *Production:* The application of *automation* to production, leads to better product quality, greater quality control and faster production. The further development of automation in the form of the *flexible workshop* is particularly useful when the production line is changed or when component machines in the production line are put out of action. Programmable automation, in the form of the robot, can be used in industry because its environment can be predetermined and its movements preprogrammed. Then, the robot has similar uses to those of the

numerically controlled machine tool, but the fields of application are different.

2. *Hazardous work:* The robot can be used in applications that would be dangerous for a human operator – high temperature applications (eg in the foundry), handling of radioactive fuels in nuclear power stations, underwater applications and others.

3. *Individual help:* The robot can be used to perform tasks that are boring, tiring or dangerous for the human operator, or, in the case of the handicapped, provide compensation for the effects of paralysis or amputation.

In the last two fields of application the robot has to deal with a variable and even unknown environment. In such circumstances the programmable robot is useless unless man retains virtual control, as in *teleoperation processes.* To deal with such problems, the robot may need to be equipped with perception and reasoning. This is one goal of current research.

Various research projects have aimed to cross the gap between such *adaptive* robots and those presently available.

Other areas of research include the development of the flexible workshop and teleoperation, both of which present considerable problems.

The present state of robotics

In Volume 1 we were concerned with the general principles underlying the functioning and definition of the robot.

So far as robot function is concerned, its level of operation is directly related to the nature and quality of the information that is introduced into the control computer in the *teaching phase.* The two main types of robot are:

1. Those that accomplish tasks, after teaching, without the use of external information introduced 'in-line'. Such 'blind' robots have 'open-loop' behaviour with respect to their invironment – everything must be known in advance.

2. Those which, following the teaching phase, rely on the use of information supplied by the environment – the tasks are executed in adaptive mode. These are the main subject of this volume.

Some non-interactive robots use artificial intelligence, for example they react to oral instructions, but would still not be classified as advanced robots as we define them, because they do not gather their own information.

Two categories of advanced robot can be recognized. There are those that can understand a problem, resolve it and apply an appropriate solution, starting with incomplete information. Then there are those that can use on-line information derived from the environment to modify their behaviour in order to attain set objectives. Current industrial robots do not fulfil either of these criteria. They can only reproduce a series of taught gestures or trajectories, in the same order as they are taught, and the only variable is that of trajectory speed. In addition, the absence of environmental sensors rules out even the slightest adaptivity by the robot to the environment.

Current research trends

Beyond the programmable automation robot described in Volume 1, research is developing in four main directions:

The first development is with robots that only use a single motor and therefore only one degree of freedom, but are capable of generating, for the end effector, trajectories more complex than the straight line or the circle. These systems, through the action of extensible components, can cover entire families of similar trajectories, through the use of only one control strategy.

The second development is concerned with an improvement in precision, speed and learning capacity over the kind of robot described in Volume 1.

The third development is concerned with the introduction of at least limited on-line adaptivity. This does not necessarily imply the use of artificial intelligence techniques. It is necessary when, whilst there is a fixed final task objective, the processes that have to be gone through to get to it may vary. For example, suppose that a machine operator has to position differently shaped pieces, arriving at a conveyor belt, on a machine. The final arrangement of each piece can be reproduced, but the sequence for the arrival of each piece and its orientation may vary. In such a situation on-line information on distance, orientation etc is needed for the final task to be completed.

The fourth (future) development is related to the autonomous power of decision in the robot — its capacity to initiate action and to learn from external conditions. The external environment might contain a man who wished to 'speak' to the machine directly. In this context, the replacement of suitable devices, such as teleprinters, that can easily communicate with the computer, presents considerable problems. Among these are the translation of ordinary acoustic signals into appropriately coded signals, involving analysis and word recognition techniques. Also, there is the problem of the signal. The appropriate command can be given in many different ways in human speech. The robot must be able to 'perceive' its environment and theoretically it must possess sensors that endow it with *artificial senses.*

What kinds of sensors: The nature of the tasks to be completed and safety requirements of the environment would determine the nature of of the sensors used. For complete versatility, generalized sensors would be needed. In particular, touch and vision are the senses that most need to be simulated, not forgetting a processing capacity needed to analyse complex images, optical or otherwise.

Pattern recognition techniques: The image contains essential information that must be extracted and interpreted – information that would relate to the target task and is handled by pattern recognition techniques – complex processes not yet completely mastered.

Artificial intelligence: The capacity to make decisions in a robot is related to available artificial intelligence – a line of research that is progressing slowly – not having reached, as yet, the speed and reliability demanded by industrial tasks.

The improved 'simple' robot: In the second line of approach mentioned above, steady improvement of the existing robot is pursued. The simple objective is to provide an easily programmable robot which executes tasks as quickly and precisely as possible.

In the learning mode for such simple robots the keyboards are especially equippped for particular functions. For example, the function might require the end effector to move in a direction parallel with itself, or always aim at the same point in space. If the keyboard is a teleprinter, the simplest and most effective programming language would be sought. In a teaching operation, the number of stored robot positions would be minimized, and the robot would then, by using these as anchors and by minimizing various criteria, reconstitute trajectories between them, on its own.

Another improvement in the simple robot would be to try to obtain *dynamic control* ie try to pursue a trajectory of specified speed and position or to generate acceleration, speed and position control that allows the robot to go from one configuration to another in a specified time, whilst respecting the limiting conditions. These would include precise arrival in a specified position with no oscillation or overshoot.

Interaction with the environment

The basic principle of automation is to regulate a system by comparing its starting condition with its final condition, and use the comparison information to guide the robot to completion of the set task.

So the robot is piloted by comparing the values of the parameters at any given time with the parameters required. In traditional automation the parameters are clearly defined because the action is broken down into a number of easily accessible elementary functions. The action of a machine tool will comprise a number of easily specified parameters, such as motor speeds and the like.

The need for task description

In the robot the infinite number of ways a job can be executed, and the versatility that generally exists, means that the range of possibilities must be reassessed in going from one task to another. Because digital computers are used, task descriptions must be developed, and in the absence of a universal task description, they must be grouped into categories. The task description depends on the *a priori* strategy that is applied. If, for example one considers a task where one piece A is being inserted into piece B, one can either choose to specify an exact trajectory that will bring this about or one could monitor the contact stress between the two pieces and try to contrive the trajectory in such a way that it is a minimum.

A number of other descriptions and strategies could be contrived, and the lack of a universal task description is evident.

Here, the problem of control is approached by using *environmental sensors*, which provide information on the instantaneous values of task parameters. The two senses that are employed to produce the necessary information are touch and vision, as will be seen later in this volume.

Use of artificial intelligence

Existing robots under development that use AI techniques are of two principal types:

1. 'Logical' robots, which exist only as simulations of the organization and functioning of an advanced robot 'brain'.
2. 'Complete' robots that have physical robot bodies as well as artificial senses.

There are a number of logical robots in existence in various laboratories, but there are very few complete robots. They are very expensive and as yet unused in large-scale production.*

* Four 'complete' robots are: *Shakey* (1968–1973) built at Stanford Research Institute. This was a mobile robot equipped with a TV camera and proximity detectors. It received its orders from a computer, by radio. Its perceptible universe was confined to rooms with doors through which it could pass, blocks that it could move by pushing, sloping surfaces and switches. The orders were typed on a peripheral teleprinter of the control computer, in simplified English. *Jason* (1970–1975) was built at Berkeley and, like Shakey, received orders by radio from a computer. It only understood FORTRAN and LISP. Its senses consisted of a TV camera, contact feelers and infrared and ultrasonic proximeters. Jason could move, and was equipped with a manipulator arm and a word synthesizer. It provided a test-bed for two plan-generators functioning in a determined and a non-determined universe. *The exploration robot of the planet Mars*, at the Jet Propulsion Laboratory, was abandoned in 1980 for financial reasons. It concerned a 'truth operation' for the exploration of the red planet. A mobile platform supported a manipulator arm (to pick up stones), a laser telemeter and a stereoscopic camera. It was designed to analyse the general and then the local scene (examination of possible samples), navigate, move and manipulate. The robot possessed plan generators to act amd move according to instructions sent from Earth. It understood FORTRAN, SAIL and CSA. *Hilare* is a mobile, French robot inspired by Jason, sited at the Laboratory of Automatic Analysis of

Conclusions

The four main directions of research and development that will take the robot from its present non-adaptive phase have been outlined. These are all interconnected and there is a common ingredient of artificial intelligence in each of them. In practice, the problem is to find the most appropriate tool for the job, without forgetting the overall objective, namely, to find a universal solution.

In this volume we shall be concerned primarily with the third category of development mentioned – environment-robot interaction. In reality, there are environmental constraints, but even under these restrictions, there are open choices and no optimal control has yet been discovered.

Systems (CAAS) at Toulouse since 1977. It is primarily a navigation research device. It possesses a TV camera connected with a laser range finder – to 'see' in relief, infrared proximity detectors and ultrasonic detectors. It possesses three levels of decision: (a) preprocessing of gathered information and checking of movements by microprocessors; (b) organization, followed by tasks of navigation and supervision of communications by microcomputer and (c) extension of computing power by means of a radio link with a large computer.

Chapter 1

Characteristics of robots adaptable to environmental variations

Today's industrial robot is little more than a machine that can be programmed, and in this sense is not much different from a machine tool, although it has a wider range of possible applications than other machines. The next generation of robots needs to be made capable of self-correcting behaviour, and to be able to react to unforeseen changes in the environment. In this volume the characteristics of such *adaptive* robots will be considered, but first a short résumé of basic definitions used in robotics will be given.

1.1 Definition of 'robot'

In Volume 1 (Chapter 1) the robot was defined as an artificial machine with the following two characteristics:

- *Versatility* ie the capability to carry out varied actions in real space. For this, the robot needs a mechanical structure with variable geometry.
- *Self-adaptation to the environment* ie the capability of self-modifying behaviour in order to attain the operational objective, despite changes in the environment during the execution of a task.

The robot in operation can be symbolically described by four interconnected elements as shown in Figure 1.1, but today's industrial robot cannot cope with one of these elements shown ie the environment. It is this element that must be serviced in the future development of robotics.

1.2 Definition and characteristics of an articulated robot

The mechanical part of an articulated robot has a chain structure. Figures 1.2a and 1.2b illustrate this point. Each articulation, rotational or translational, is actuated by an electric, hydraulic or pneumatic motor. To describe the action of such a robot four main elements must be taken into account:

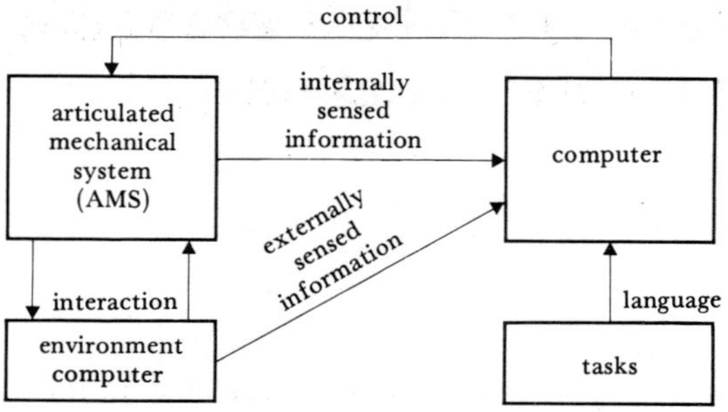

Figure 1.1. *Representation of robot function*

(a) *The task space* (R_0) or work area. This is real space. The position and direction of the robot's end effector are described in this space by vector $\underline{X}(R_0)$ with six components, of which, more often than not, three describe the position of a particular point on the tool (for example its centre of gravity) and three its direction. The components of $\underline{X}(R_0)$ are sometimes called *operational coordinates.*

(b) *The articular variables* in which the robot's configuration is denoted by vector $\underline{\Theta}$. This has as many components as degrees of freedom (in most cases). The components of this vector are called *generalized coordinates*.

(c) *The torques* applied to articulations that are expressed by vector $\underline{C}$.

(d) *The control element* in which the real control vectors (voltages or currents applied to the inputs of the motor slave units, for example) are written as $\underline{V}$.

Robot control can be encapsulated in the following equation, describing direct and inverse flow:

$$\underline{V}(t) \rightleftarrows \underline{C}(t) \rightleftarrows \underline{\Theta}(t) \rightleftarrows X(t)R_0 \quad (1\text{-}1)$$

and Volume 1 is essentially devoted to this equation. In Volume 2 means of automatically generating the operational coordinates $\underline{X}(t)(R_0)$ in a partly unknown environment, subject to certain restrictive hypotheses, are sought.

The computer starts with a model of the articulated mechanical system, a model of the environment, data related to the tasks to be carried out and a certain number of algorithmic resources (strategy and calculation). In operation, the computer receives, in addition, information on the present state of the articulated mechanical system (internally sensed information) and on the present state of the environment (externally

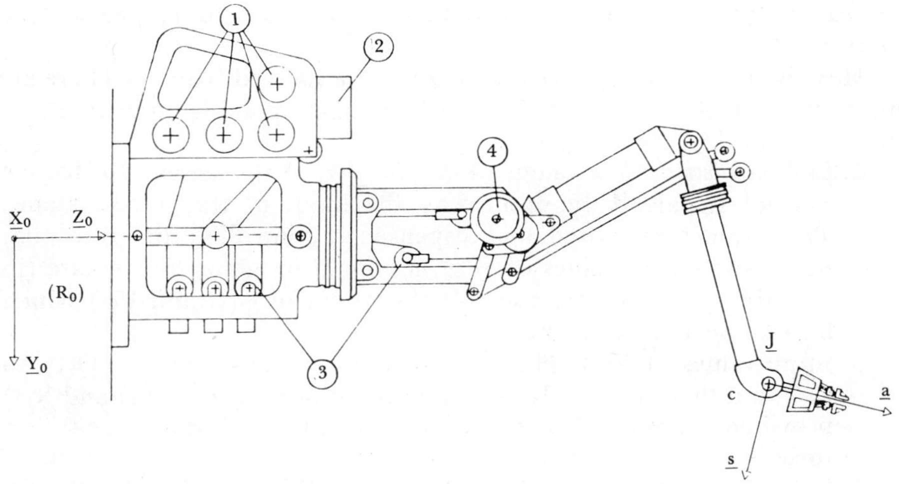

Figure 1.2a. *The robot manipulator MA23 (CEA, La Calhene). 1. Motors; 2. servo-systems; 3. transmission units; 4. counter balance*

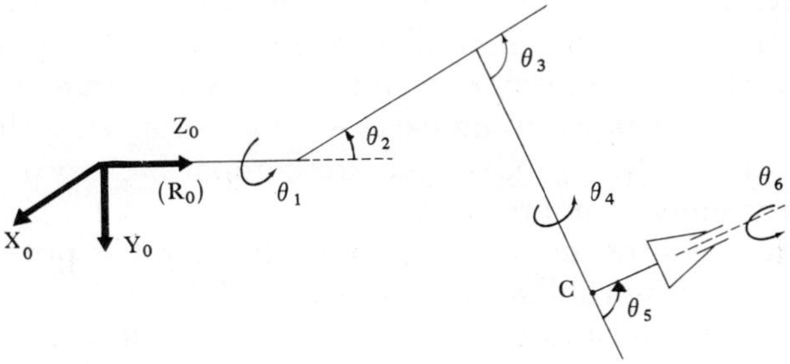

Figure 1.2b. *Geometrical structure of the robot manipulator shown in Figure 1.2a. The first three DOF (θ_1, θ_2, θ_3) allow point C (ie the gripper) to be positioned in space. The last three DOF (θ_4, θ_5, θ_6) allow the gripper to be oriented in any direction*

sensed information). With this information and with appropriate resources it calculates a form of control to be applied to the actuating motors of the articulated mechanical system, which can thus complete the task assigned to it.

1.3 Robot programming

Suppose that equation (1-1) comes under perfect computer control for a given robot, as defined in Volume 1. The object is to enable this robot to carry out a sequence of movements or torques developed by the motors.

How is the robot to be informed what is expected from it? There are numerous possibilities of which the following program is typical:

1. Estimate successive values of vector $\underline{V}(t)$ ie the series of effective control signals to be applied to the input of the servo-systems. These values are stored, and dispensed at controlled time intervals.
2. Store successive values of $\underline{\Theta}(t)$, leaving a program to estimate (in real time for dynamic control) the values of $\underline{C}(t)$ and $\underline{V}(t)$ which have to be applied to the robot.
3. Store values of $\underline{X}(t)$. These will have to be transformed to $\underline{\Theta}(t)$, to $\underline{C}(t)$ and then to $\underline{V}(t)$. In this case the values of $\underline{V}(t)$, $\underline{\Theta}(t)$ and $\underline{X}(t)$ provided correspond with sufficient samples to describe the trajectories, without undue loss of information.
4. Store configurations $\underline{X}(t_1)$, $\underline{X}(t_2)$ etc (Volume 1, Chapter 9), leaving the computer to interpolate between these configurations, first in its work domain, and then in its control domain. It is only in the last case that the robot shows any *capacity for initiative*, as the trajectories are incompletely specified. So far one has given the computer explicit instructions. All constraints have been stated in a mathematical form, including those concerning possible interaction with the environment. In the absence of dynamic control it suffices, in the most complex case, (and supposing that the time constraints on $\underline{X}(t_1)$, $\underline{X}(t_2)$ etc are compatible) to use an ad hoc trajectory-generation program.
5. It is not until the next stage, when the robot is provided with a description of a problem to be solved (or of a task to be executed) in a *linguistic* and not a mathematical form, that the robot has to use a further attribute: to understand the problem and solve it autonomously in order to bring about the required action. The description of the problem may not be provided in fine detail, as in communication between one person and another. Some elements remain implicit and must, therefore, be deduced. Other non-deducible elements will have to be found in the environment using an ad hoc strategy.
6. The problem becomes more complicated as the language used to communicate with the robot approaches natural speech. The most complex stage is reached when the problem is set out for the computer as in normal speech. The robot not only needs to be able to understand human language as it might be faultlessly transmitted to it by a teleprinter, but now also needs to be able to translate sound waves into phrases, word groups, syllables etc, understand their meaning and then solve the problem. So research on speech

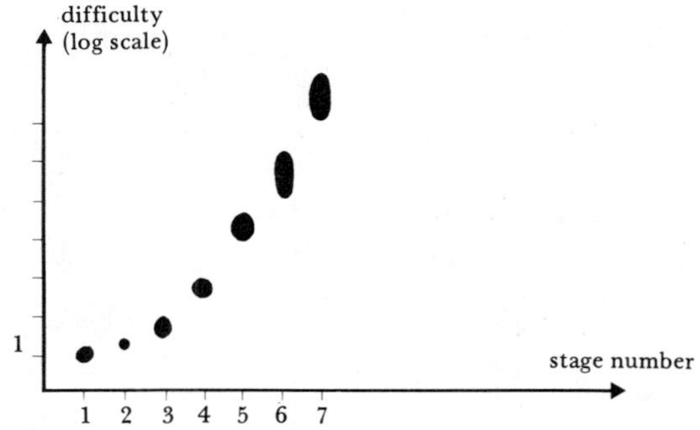

Figure 1.3 *Graph representing difficulty in mastering different stages of robot description vs number of stage*

recognition is included in the problem which needs to be solved. Only from stage 5 onwards can one speak of artificial intelligence for robots.

The growth in complexity of the problem tends to be exponential, proceeding from stage 1 to stage 7, as shown in Figure 1.3. In this volume the stage reached is stage 5 ie a task is described in a given language. Also the robot will have to carry out a task that has been incompletely specified and also to interact with information from the environment.

1.4 Structure of the problem to be solved

The general question of environment-robot interaction can be described in simple terms. Given an adequate task description, how should the robot use its resources to carry this task out effectively and correctly in a reasonable time span? The answer is, not surprisingly, far less complex than the question.

First of all, the task needs to be accomplished with positional precision and then the run-time needs to be minimized. The often-opposing needs of precision and speed of operation do not affect the reasoning involved, but do affect the control designs employed.

The conceptual structure, in which interaction between environment and robot is represented (for an industrial robot with fixed position), is shown in Figure 1.4. It consists of three main parts:

- Understanding the problem.
- Assessment of available information and equipment before starting the work.

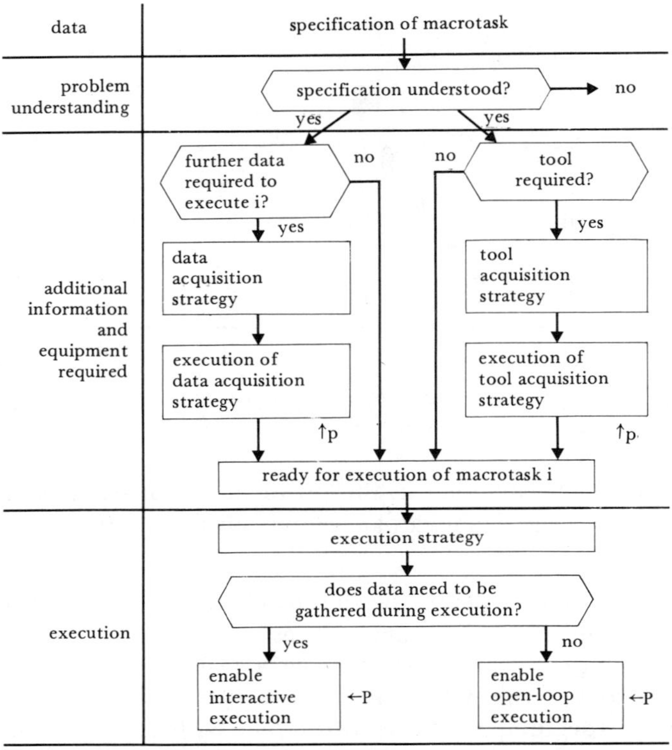

Figure 1.4 *Flow diagram for adaptive macrotask execution. Arrows P indicate that, during execution, the sensors can interrupt and reset the execution strategy*

– Execution of the task, which is termed a *macrotask* because it can be made up of a large number of elementary tasks, as described in Volume 1, Chapter 10.

Each macrotask is characterized by its independence, the only link between individual macrotasks being a possible preferred order of execution. For example, sawing a plank and driving a nail into the ceiling make up two macrotasks whose independence is diminished only if the plank is used to stand on to put the nail in its place. For purposes of simplicity, the acquisition of the tool as well as understanding of problems which specifically call for techniques of artificial intelligence will be left to one side. The question 'Should the information be acquired before the execution of task i?' depends considerably on the data provided at the entry ie description to the robot of macrotask i. This is why a 'complete' and 'incomplete' description of the macrotask needs to be defined.

1.5 Description of the macrotask – complete and partial

To define what is meant by a complete description, start with an example. Suppose that the data concern the transport of object A, which is placed among other objects, to point B (Figure 1.5). Possessing these data, together with its algorithmic resources, the robot can employ any or all of the following procedures to acquire the missing data which will allow it to move into the execution stage.

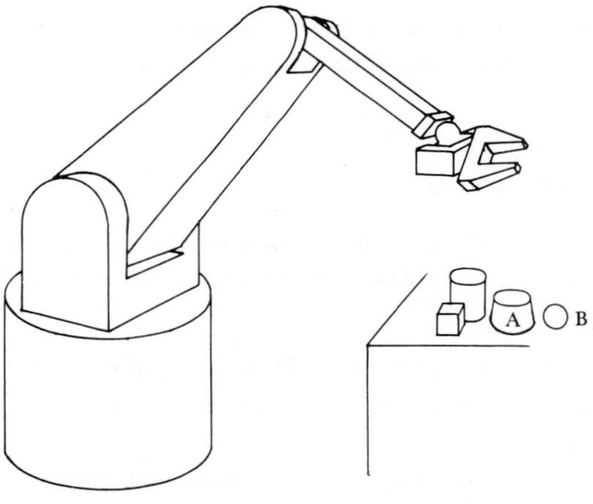

Figure 1.5 *Grasping object A followed by transport to point B*

1. It can make an *algorithmic deduction*. Suppose the robot needed to know the weight of object A, and that the description included volume and density only, then weight could be deduced easily from these facts.
2. It can *ask* the user for *data*. If the volume and density of A do not appear in any way, it can, if it needs to, ask 'What is the weight of the object?'.
3. It can *estimate* or *invent* data. For instance, if the weight has not been supplied, the robot can call on past experience or take a standard value *a priori*.
4. Finally, it can work out a *strategy of data acquisition*. For instance, if the description of the macrotask does not include the positioning of object A, it will use its sensory transducers to find it. This case will be of particular interest in what follows.

From now on, a task description will be called *complete* if all elements necessary for the start of its execution are present in the description or are deducible using algorithms (without supplementary data or information). As shown in Figure 1.4, this means that one can go directly from the 'data' level to the 'ready to execute the macrotask' level. That is not to say that there will be no need to collect information during the 'execution' stage.

1.6 Review of intended task description

In Volume 1, Chapter 10 an inductive description of macrotask was suggested. In this, from a knowledge of the robot, one arrives at its control through easy steps. Before applying it to the problem stated in section 1.4, let us recall its elements. The macrotask is described at four levels:

1.6.1 THE TASK GOAL (TAG)

This is independent of the robot, and refers to the difference between the start and end states of the robot's environment. This is described by three subgroups:

(a) The objects of interest (OI) with their initial state (IS), their transitory states (TS), their normal state (NS) and their final state (FS).
(b) The characteristics of the environmental medium (EM).
(c) Other objects (OO).

1.6.2 THE TASK TOOL (TT)

This comprises a description of the procedure needed to achieve the task goal, the tools needed and their working rules.

1.6.3 THE INTERACTION TASK (ITA)

This comprises elements not present in TAG and TT but which necessarily have to be mastered and which arise from the interaction of the robot with the environment.

1.6.4 THE ROBOT TASK (ROT)

This consists of (starting with the TAG, TT and ITA groups) the creation of first in the task space, then in the space of articular variables and finally in that of the control variables, those trajectories which the robot must perform.

1.7 Structure of the problem to be solved using task description

In every description of tasks or macrotasks (abbreviated DMT) an ambiguity often missed by designers exists: the confusion of description and usage. Or to put it more accurately, a description should arise from desired usage. Thus the DMT described in section 1.6 in the form of four sets is not related to the task in abstract only but follows from the intended execution by an articulated robot. In reality, the macrotask is totally included in TAG. The taking into account of necessary elements for its execution by a robot comes in TT and ITA. The method which the robot is going to use to succeed in the task execution, supposing that the characteristics of the robot are known, is condensed in ROT. From then on, the flow diagram in Figure 1.4 can be redrawn as in Figure 1.6, using the DMT. The following two points must then be taken into account:

1. Knowledge of the characteristics of the robot and of TAG, TT and ITA which allow ROT to be induced.
2. Clear distinction must be drawn between TAG and its preparation

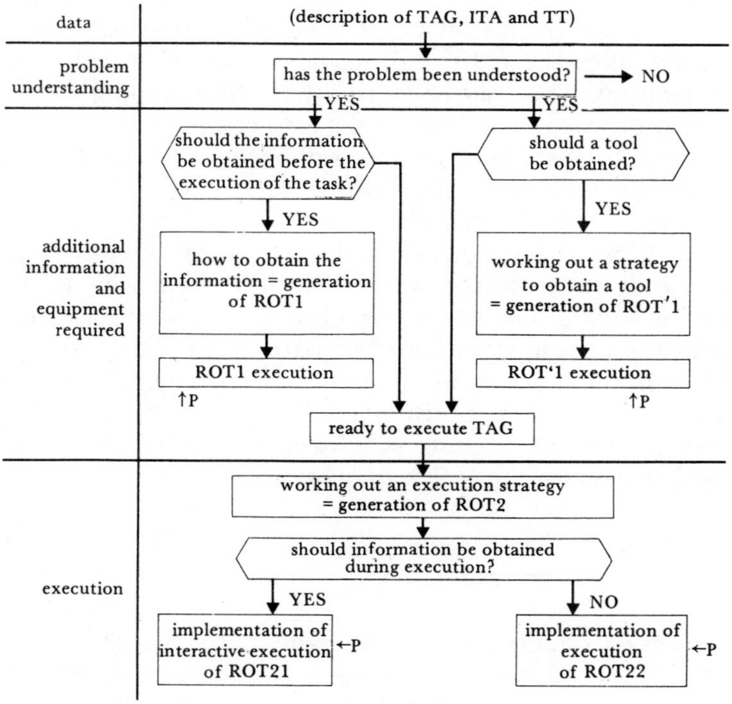

Figure 1.6 *Execution of a macrotask using the DMT as described in section 1.6 and assuming the availability of various resources*

ie the compilation of information. For example, if TAG describes the transport of object A to point B, the information-gathering phase (which is not part of TAG), will consist of obtaining the position of A in the line linked with the robot.

1.8 Environment-robot interaction

The robot is plunged into a physical universe, and one has to take into account the physical laws governing this universe. The nature of the medium in which the robot is immersed and its effect on the movement of the robot must first be considered. The medium might be a common gas like air or a rare gas such as argon; each will present resistance (viscosity) to movement. In underwater applications, the medium will almost certainly be water, whose resistance is a function of depth. Then there are gravitational effects due to the attraction that exists between masses. For robots in outer space the gravitational effect can come from the moon or the sun or any other celestial body in whose proximity the robot is travelling.

Moreover, the robot is going to have to move whilst carrying an *effective load* (Volume 1, Chapter 10), which can be the object to be handled or a tool which is changed from one macrotask to the other. This load modifies the mass and inertia of the robot and affects possible acceleration. The load will also affect the clamping force needed to keep it in position. The bulk of the load can prevent the robot from reaching the configuration it might assume when unloaded. Clearly, the weight of the load too, has a strong effect on behaviour. Even unloaded, the robot moves in space that contains other solid bodies: objects to be transported as well as various obstacles. If the robot end effector, for example, needs to cross the work table, the table will itself oppose movement as soon as *contact* is established. Forces of reaction will be set up which are going to modify the state of the robot. Finally, *interaction at a distance* occurs when the robot is equipped with arbitrary sensors. Recognition of the position and direction of an object, through, for example, a camera, is going to influence the behaviour of the robot if its hand has to grasp the object in question.

1.9 Interaction and constraints

Let us take the example again of the manipulator that tries to cross the work table. The table-robot interaction is only a potential one while there is no contact. One can have two attitudes with regard to this interaction:

(a) Either one wants to *avoid* it and gives expression to this as a geometric constraint. There will be 'forbidden zones' through which the robot is not allowed to pass.
(b) Or one can *control* the nature or extent of the interaction: for example the robot can be prevented from pressing on the table with a force of more than 20 newtons, or made to wipe the table with a cloth whilst permanently exercising a vertical force of 10 newtons.

It must be remembered that these two attitudes are very different in practice from an open-doors execution plan of the macrotask that might be assigned to the robot.

In the first case, if constraint is respected, movement is free ie without contact with the environment. If a few safety factors are created for the forbidden zones, the only remaining effect that might come into play is that the table acts as an obstacle. Ignorance of this may not be serious.

In the second case there is physical contact between the robot and the environment. One can say that the robot 'feels its way around' the environment. The table becomes an object of interest and every mistake can be serious. One then evokes concepts of *strong interactions* and *weak interactions*, where a strong interaction cannot be tolerated whilst a weak interaction can be, given that its effect can be measured.

The interactions which qualify as 'strong' can be classified into two categories:

1. Those produced by the surroundings in which the robot moves:
   (a) Interaction with the gravitational field, which constrains the acceleration of the robot.
   (b) Interaction with the medium (air, water etc). This introduces viscous forces of friction or resistance against movement and thus constrains attainable speeds.
2. Those produced by the objects present in the robot's range:
   (a) Interaction produced by the transported load. The load can limit attainable acceleration and configuration.
   (b) Interaction produced by contact with obstacles ie the objects in the surroundings which only need to be avoided. As already seen, these interactions are going to induce geometric constraints (forbidden zones) or tolerance constraints on the intensity of interaction between the obstacle and a part of the robot.
   (c) Interactions produced by contact with the object of interest (for tactile recognition, stress inspection, gripping.). Unlike (b) we have some knowledge of these interactions in the sense that they can often be expressed as equations with a solution. They

naturally induce geometric constraints and stress limits. They also induce contraints which could be called strategic ie those which have been deduced from a reasoning of the kind 'Knowing that I have just obtained new information or have just carried out part of the macrotask, how am I going to pursue the execution of this macrotask?'. Note that strategic constraints can also come from interaction with obstacles.

'Weak' interactions are defined as all those that are formed without physical contact with the environment. They only induce primary strategic constraints (themselves generators of secondary constraints on numerous factors). The notion of strong and weak interaction does not affect the complexity of the problem to be solved. It only provides a measure of magnitude for allowable errors.

1.10 Ways of dealing with interaction

This point is very important, for it is directly related to the ability of the robot to initiate action and therefore to its degree of development.

Modelling the interaction is in fact modelling for the effect of the environment on the robot and foreseeing the robot's reaction. This reaction is expressed at two levels. One is *passive:* the robot 'feels' the action by noting that its state (ie a group of parameters representing it) is being modified under this action. The other is *active:* the variation of the state is translated (interpreted) into decisions and commands for movement (for example). Looking at it from another, more important, angle, one could say that every effect of the environment on the robot expresses itself through force(s).

But these forces can be *direct* (examples: action of gravity, obstacles which create resistance to the robot's passing etc) or *indirect* (examples: object in the field of a camera bound to the wrist of the robot which stops the robot when in view; automatic positioning of the gripper to grasp an object with the help of proximity detectors etc). In these conditions, one finds that:

- *Either* these forces, or the elements which allow them to be calculated, are known in advance. The interaction can therefore be taken into account before the robot executes its task. This is always the case for interactions arising from gravity or the surrounding medium.
- *Or* these forces and the elements which allow them to be calculated can only be obtained during execution. This case always arises during *contact* between the robot and the objects of interest in movement, stress inspection or gripping modes.
- *Or* one has a *combination* of the two preceding cases. Thus the

		Physical constraints			*Strategic constraints*		*Method of assessing interaction*
Type of interaction	Origin of interaction	Geometric constraints	Velocity constraints	Acceleration constraints	Constraints on forces developed	Software constraints	
Strong interaction	Gravity			X			*a priori*
	Fluid or surrounding material		X				*a priori*
	Transported load	X		X			*a priori* or on-line
	Contacts with obstacles or limits	X			X	X	*a priori* or on-line
	Contacts with the objects of interest (palpitation – stress – controlling) gripping	X			X	X	on-line
Weak interaction	Identification at a distance					X	*a priori* or on-line

Figure 1.7 *Characterization of the environment-robot interactions*

transported load may or may not be known beforehand. The same is true for the obstacle, the nature, the position and the direction of the objects of interest. For each type of interaction, the nature of the primary constraints and the ways of taking them into account are summarized in Figure 1.7.

1.11 Environment-robot interaction and artificial senses

To react appropriately to the effect of the environment, the robot must sense and interpret it. If there is physical contact or if the robot is in a gravitational field, it senses this effect, since its mechanical body is sensitive to it, but without extra sensing it could damage itself or the environment. If it is in a field of non-gravitational forces it is not aware of anything.

The robot must therefore be equipped with sensors that measure interaction, and must also be equipped with a 'brain' to allow interpretation of the measurement. These sensors are, by analogy with living beings, called 'artificial sensors'.

The next question relates to the nature of the artificial senses required by the robot. The answer obviously depends on the macrotasks that the robot has to carry out and even more on the information on the interactions that will be supplied to the robot in describing the task. If the description is complete and execution is in open loop with regard to the environment, external artificial senses are not needed. On the other hand, if the robot has to perform an incompletely specified task in a partially unknown environment, external sensors might have to be varied, numerous and complex. This question cannot be answered mathematically because the final robot can have such varied abilities.

Therefore one turns again to the example of the human being: a perfect robot who has, to a level of perfection, the characteristics of versatility and adaptativity which we seek to imitate. At first sight, man has five well known senses: vision, touch, hearing, smell and taste. A deeper study shows that the surface of the human body is literally covered with very diverse environmental sensors: heat sensors, pain sensors, motor receptors of various kinds.[2] In all, there are more than 200 different types of environmental sensors available to man. Note that a confusion has been allowed to establish itself between 'sense' and 'external sensors'. Man's five senses are global functions resulting from the integration of information from numerous sensors. In order to see we have two eyes, but in a particular case of a blind person it has been noted that touch sensors in the fingertips can develop a perception comparable with vision.[3]

So which of the five sensory functions of man seem necessary or useful to the robot? To answer this, one must look at the role of each of these functions in human behaviour. Smell and taste are senses which effectively act as agents of chemical analysis of the environment. For the robot in the workshop they are of little interest. A man who has become deaf can indulge in many activities: his handicap is related to oral communication; he is at risk because he cannot detect noises warning him of danger. But our robot is not advanced enough to understand oral commands, so let us eliminate this sense provisionally. Vision and touch are left. These are the most important senses needed for interaction with, and modification of, the environment. It is these senses that will be studied with regard to the articulated robot.*

* For a robot other artificial senses can be created for *particular* applications: sensitivity to an electric field, magnetic field, radioactivity etc.

1.12 Artificial senses and redundancy of information

Let us compare the strategies of a sighted man and a blind man in the street. The former recognizes his route, mainly with his eyes, but also through his hearing and with the reaction forces between the ground and his feet. The blind man, however, makes use of the same information, other than sight, as well as that obtained by means of a stick with which he senses the ground. Both these two men can complete the same task eg can walk up a particular street on the pavement on the left side of the street. This shows that the same set of sensory functions can be used in different ways to carry out a given task. For man, the suppression of an apparently essential sense like vision does not leave him totally unequipped to carry out a large number of tasks. For a given task, given identical resources, the quantity of information needed from the environment is constant. Only the method by which information is acquired varies. In our example the same quantity of information can be obtained (to take either man safely up the street) with sight, hearing and the tactile senses (with the sighted man) as with hearing and the tactile sense (with the blind man). So there is a redundancy of information which vision can supply and which hearing and touch can supply.

Redundancy between vision and touch is particularly noticeable when used by man in carrying out motor tasks. However, exploitation of such redundancy is not natural in man, who, in fact, selects the sensor which is most suitable for a particular action. This is why one talks of *replacement sense* or *compensation sense* when describing tactile sense in a blind person. The problem of redundancy of sensory functions and of compensation sense is relevant to advanced robots: an object can be found by performing to the work plan using an end effector equipped with tactile sensors, or by using a camera. When redundancy is present in a robot, it can be used in two interesting ways:

(a) It allows, by 'cross-checking', improvement in the quality of the information obtained by a sensor or a group of sensors. This enhances the sensory function.

(b) It also allows for a replacement strategy when a system which supplies redundant information becomes faulty. The reliability of the robot is increased by it.

Figure 1.8 is a symbolic representation of the external information necessary for the execution of a macrotask. It will be noted that:

(a) A sensory function can be supplied by a group or groups of sensors with partially redundant information. I_1 is redundant with respect to parts of I_2 and I_3. I_4 is redundant with respect to part of I_3.

(b) The various artificial senses are themselves partially redundant. I_7 encroaches on I_1 and I_4. I_5 encroaches on I_8.

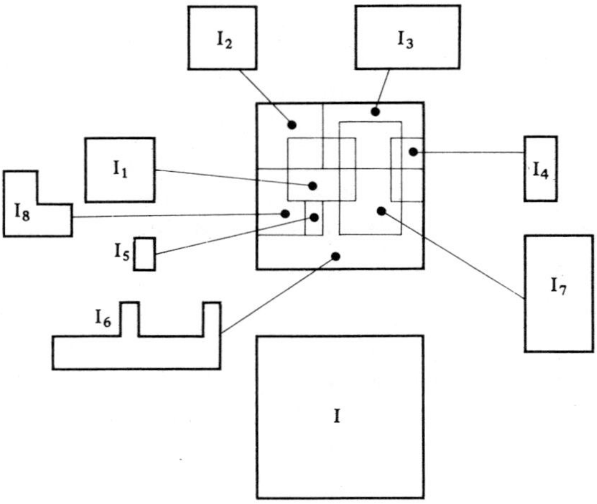

Figure 1.8. *Redundancy of information. I. Total external information necessary for the execution of a macrotask by the robot. I_1, I_2, I_3, I_4. Information which can be deduced from the groups of sensors c_1, c_2, c_3, c_4. The four sensors form a sensory function or an artificial sense, for example, vision. I_5, I_6, I_7. Information which can be deduced from the groups of sensors c_5, c_6, c_7. These three groups of sensors form another artificial sense, for example, the tactile sense. I_8. Information available in the sensors c_8 which form another sensory function eg hearing*

Conclusions

It has been argued in this chapter that a robot equipped with two primary senses (touch and vision) would advance the 'old' robots described in Volume 1 into the 'advanced robot' which is the subject of this volume. In an industrial environment, it is vital that, even if the tasks performed are not as general as have been described in this chapter, they should be performed speedily and effectively. In particular, careful thought must be given to the processing of sensory data and the generation of actions that ensue. The first step is to understand some basic sensory interactions, and this is the subject of Chapter 2.

Chapter 2

Interaction with weight, surrounding medium and workload

The effect of gravity and the surrounding media are both permanent and stationary. This is why these two factors are taken into account at the design stage of the robot and have an effect on the action of the robot controlled by a programmable machine.

2.1 Interaction with gravity

In Volume 1, Chapter 7, a dynamic model of the robot formed by an articulated chain of M rigid bodies was established. This model is described by N coupled, non-linear, differential equations of the form:

$$\sum_{\substack{j=1 \\ i=1,N}}^{N} \{ A(i,j) \cdot \ddot{\theta}_j + C(i,j)\dot{\theta}_j^2 + \sum_{k=j+1}^{N} B(i,j,k) \cdot \dot{\theta}_j \dot{\theta}_k \} = Q_i + \Gamma_{\theta i} \qquad (2\text{-}1)$$

which can be condensed to matrix form:

$$\underset{N\times N}{[A(\underline{\Theta})]}\underset{N\times 1}{\ddot{\underline{\Theta}}} + \underset{N\times C_N^2}{[B(\underline{\Theta})]}\underset{C_N^2\times 1}{\dot{\underline{\Theta}}\dot{\underline{\Theta}}} + \underset{N\times N}{[C(\underline{\Theta})]}\underset{N\times 1}{\dot{\underline{\Theta}}^2} = \underset{N\times 1}{\underline{Q}(\underline{\Theta})} + \underset{N\times 1}{\underline{\Gamma}_\theta} \qquad (2\text{-}2)$$

In equation (2-2)

$$\underline{\Theta} = (\theta_1, \theta_2 \ldots \theta_N)^T$$

is the vector of the generalized variables (the most frequently found articulated variables) and

$$\dot{\underline{\Theta}}\dot{\underline{\Theta}} = (\dot{\theta}_1 \dot{\theta}_2, \dot{\theta}_1 \dot{\theta}_3, \ldots \dot{\theta}_1 \dot{\theta}_N \quad \dot{\theta}_2 \dot{\theta}_3, \ldots \dot{\theta}_2 \dot{\theta}_N, \ldots \dot{\theta}_{N-1} \dot{\theta}_N)^T$$

This vector comprises

$$C_N^2 = N!/2!N-2!$$

$$\dot{\underline{\Theta}}^2 = (\dot{\theta}_1^2, \dot{\theta}_2^2, \ldots \dot{\theta}_N^2)^T$$

The term $[A(\underline{\Theta})]\ddot{\underline{\Theta}}$ in equation (2-2) corresponds to the inertial forces; $[B(\underline{\Theta})]\dot{\underline{\Theta}}\dot{\underline{\Theta}}$ with Coriolis forces and $[C(\underline{\Theta})]\dot{\underline{\Theta}}^2$ with the centrifugal forces. The term $\underline{\Gamma}_\theta = (\Gamma_{\theta_1}, \Gamma_{\theta_2}, \ldots \Gamma_{\theta_N})^T$ is the vector representing active couples, ie coming from the actuators and exerting an effect on the articulations.

$\underline{Q}(\underline{\Theta})$ is the vector representing couples influencing the generalized variables arising from the effect of gravity. It represents the influence of gravity on the robot.

The value of component i of vector $\underline{Q}$ is expressed in the form:

$$Q_i = \sum_{\lambda=1}^{M} gM_\lambda \frac{\partial Y_{G\lambda}^{(0)}}{\partial \theta_i} = \sum_{\lambda=1}^{M} gM_\lambda f_i(\underline{\Theta}) \qquad \textbf{(2-3)}$$

where g is the acceleration due to gravity, M_λ is the mass of body λ, $Y_{G\lambda}^{(0)}$ is the height of the centre of gravity G_λ of the body λ relative to the fixed line R_0 connected with the robot.

Example: Consider the arm in Figure 2.1, which has three rotational DOF. Let us consider the couples created by the effect of gravity on the three articulations $\theta_1, \theta_2, \theta_3$.

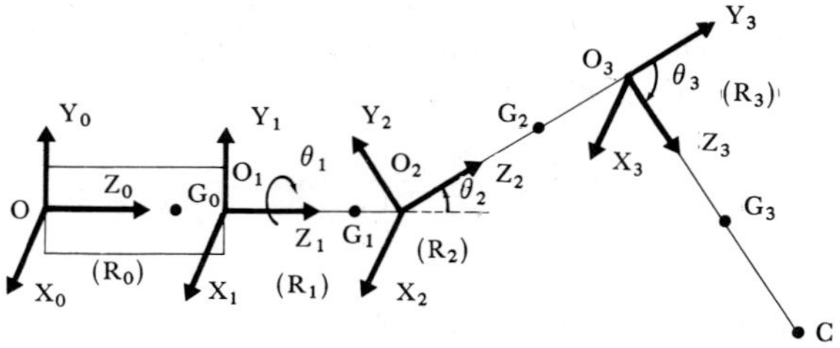

Figure 2.1. *Arm and associated transformation*
$O_1 G_1 = \ell_1; O_2 G_2 = \ell_2; O_3 G_3 = \ell_3$
$OO_1 = L_0; O_1 O_2 = L_1; O_2 O_3 = L_2; O_3 C = L_3$

Referring to Volume 1, the matrices of the changes are written:

Movement from R_1 to R_0 through a rotation θ_1 about Z_0:

$$M_0^1 = \begin{pmatrix} C1 & -S1 & 0 \\ S1 & C1 & 0 \\ 0 & 0 & 1 \end{pmatrix} \quad \text{where } C1 \equiv \cos\theta_1,\ S1 \equiv \sin\theta_1 \qquad \textbf{(2-4)}$$

Passage from R_2 to R_1 through a rotation θ_2 about X_1:

$$M_1^2 = \begin{pmatrix} 1 & 0 & 0 \\ 0 & C2 & -S2 \\ 0 & S2 & C2 \end{pmatrix} \quad \text{where } C2 \equiv \cos\theta_2,\ S2 \equiv \sin\theta_2 \qquad \textbf{(2-5)}$$

Passage from R_3 to R_2 through a rotation θ_3 about X_2:

$$M_2^3 = \begin{pmatrix} 1 & 0 & 0 \\ 0 & C3 & -S3 \\ 0 & S3 & C3 \end{pmatrix} \quad \text{where} \quad \begin{matrix} C3 \equiv \cos\theta_3, \\ S3 \equiv \sin\theta_3 \end{matrix} \tag{2-6}$$

Writing the vectorial relations as:

$$\underline{OG_1}(R_0) = \underline{OO_1}(R_0) + \underline{O_1G_1}(R_0) \tag{2-7}$$

$$\underline{O_1G_1}(R_0) = M_0^1 \underline{O_1G_1}(R_1) \tag{2-8}$$

$$\underline{OG_2}(R_0) = \underline{OO_1}(R_0) + M_0^1 \underline{O_1O_2}(R_1) + M_0^1 M_1^2 \underline{O_2G_2}(R_2) \tag{2-9}$$

$$\underline{OG_3}(R_0) = \underline{OO_1}(R_0) + M_0^1 \underline{O_1O_2}(R_1) + M_0^1 M_1^2 \underline{O_2O_3}(R_2) + M_0^1 M_1^2 M_2^3 \underline{O_3G_3}(R_3) \tag{2-10}$$

the heights of the centres of gravity of the different bodies are obtained:

$$Y_{G_0}^{(0)} = 0; \quad Y_{G_1}^{(0)} = 0; \quad Y_{G_2}^{(0)} = -C1S2l_2 \tag{2-11}$$

$$Y_{G_3}^{(0)} = -[C1S2L_2 + C1S(2+3)l_3] \tag{2-12}$$

When applying equation (2-3) the following relationships arise:

$$Q_1 = gS1S2\,[M_2 l_2 + M_3 L_2] + gS1S(2+3)M_3 l_3 \tag{2-13}$$

$$Q_2 = -\,gC1C2\,[M_2 l_2 + M_3 L_2] - gC1C(2+3)M_3 l_3 \tag{2-14}$$

$$Q_3 = -\,gC1C(2+3)M_3 l_3 \tag{2-15}$$

The couples developed by gravity result from the configuration $\underline{\Theta} = (\theta_1, \theta_2, \theta_3)$ in a totally non-linear way. Suppose the bearer has stopped. This would mean that the actuators develop couples Γ_{θ_i} that work against couples Q_i:

$$\Gamma_{\theta_i} = -\,Q_i \tag{2-16}$$

For this equation to be satisfied, the actuators should always be active, because gravity is a permanent factor.

In practice this would not be convenient. Instead, two methods are used to prevent the robot from crashing to the ground:

1. Links between actuators and articulations which are irreversible (the equivalent of a ratchet system) are used.
2. Mobile masses which provide an invariant balance for a large group of configurations are added. The parallelogram technique is the one most often used (Figure 2.2). In these conditions the influence of gravity can be significantly reduced at the cost of increase in

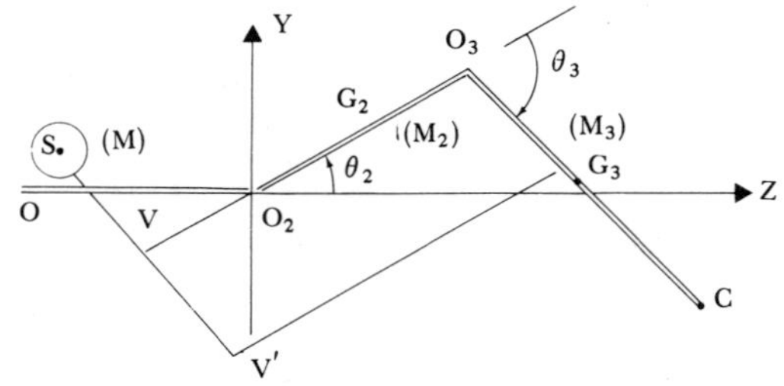

Figure 2.2. *Example of arm equilibrium (in Figure 2.1) for two DOF. IF $VS = M_3 O_3/2M$; $O_2 V = M_2 \cdot O_2 O_3/2M = M_3 \cdot O_2/O_3/2M$, the couples with articulations O_2 and O_3, produced by gravity, become zero whatever the position of C if the bearer moves in plane YOZ*

mass and inertia of the robot. Note that in practice the couples produced by the arm can be balanced out but that it is quite often mechanically impossible to balance out those due to the end effector, in particular, the forces due to the transported load if it has three DOF.

2.1.1 ACTION OF GRAVITY AND CONTROL

In Chapter 1 interactions taken into account *a priori* and on-line were discussed. It was also seen that gravity may be taken into account *a priori*. This obviously means that, knowing the planned trajectory beforehand, couples exerted by gravity at each point of the trajectory can be calculated in advance. This must not be confused with the implementation of a practical trajectory, which would call for permanent inspection of the values of the different variables involved. For example, supposing it was calculated in advance that, at time t_i, the variables θ_i and $\dot{\theta}_1$ should have the values of θ_{t_i} and $\dot{\theta}_{t_i}$. If they do not correspond with the required values, the mistake allows use of equation (2-2) to generate a correction. In these conditions the calculation must relate to the situation in line with the vector $\underline{Q}$. The complexity of the mathematical expression for $\underline{Q}$ will have a direct bearing on the control performance (Volume 1). For this reason, the mechanical compensators mentioned above raise interesting problems.

2.1.2 EFFECT OF GRAVITY ON PROPOSED STRUCTURE OF THE ROBOT

Consider a satellite which is far from Earth and which carries an articulated robot. g is very small under these circumstances and all couples previously mentioned cancel each other out. The mechanical structure of the robot can be lighter because stabilizers are no longer needed.

Moreover, rapid response will not be obtained in the articulated parts, because there are problems due to delay in the transmission of the commands sent from Earth. This robot will not be used for production purposes but for maintenance, inspection, repairs etc in space. Taking these factors into account, the system described by equation (2-2) will be greatly reduced to:

$$[A(\underline{\Theta})]\,\ddot{\underline{\Theta}} = \underline{\Gamma}_\theta \tag{2-17}$$

But each couple applied to an articulation is going to induce a reaction couple on the robot + satellite, by completely modifying the robot's structure or by developing couples such that the inertia of the satellite can absorb the reaction.

2.1.3 CASE OF A NON-RIGID ROBOT

The structure of the articulated robot can be flexible in two different ways:

1. The members that form it can bend under the action of the forces which are applied to it.
2. The articulations can possess 'elasticity' arising from the effects of transmissions. This is particularly true for transmissions of the cable or belt type (Figure 2.3).

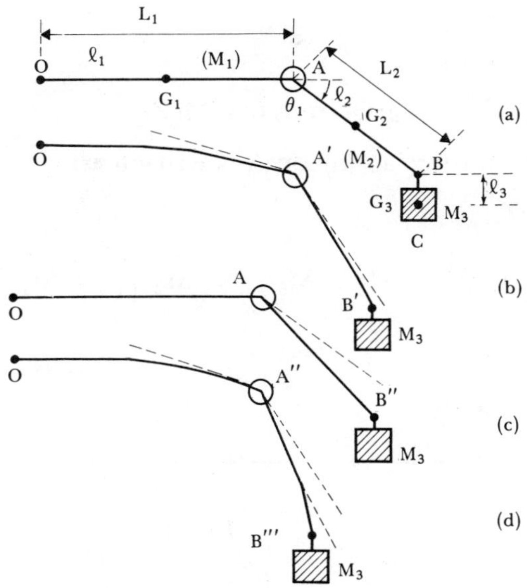

Figure 2.3. *a, rigid segments – rigid articulation; b, flexible segments – rigid articulation; c, rigid segments – flexible articulation; d, flexible segments and articulations*

At low speeds, gravity is going to create much elasticity in the segments and articulations (in practice, however, segments can nearly always be considered rigid). Consider the simplified example shown in Figure 2.3. Gravity exerts on the one hand a rotational couple on A:

$$Q_A = -gM_2 1_2 C1 - gM_3 L_2 C1 \tag{2-18}$$

and exerts on the other hand vertical forces on the structure. To calculate this force on A, which can entail a deformation of segment OA, the centres of gravity of the segments OA, AB, BC must be found. Suppose that these segments have, as centres of gravity, G_1, G_2 and G_3. Using the definition

$$M \cdot \underline{OG} = \sum_{i=1}^{3} M_i \cdot \underline{OG_i} \tag{2-19}$$

the position of G (Figure 2.4) can be easily found:

$$Z_G = \frac{M_1 1_1 + M_2 (L_1 + 1_2 C1) + M_3 (L_1 + L_2 C1)}{M_1 + M_2 + M_3} \tag{2-20}$$

$$Y_G = -\frac{M_3 L_3 + (1_2 M_2 + L_2 M_3) S1}{M_1 + M_2 + M_3} \tag{2-21}$$

and $$M = M_1 + M_2 + M_3 \tag{2-22}$$

where $C1 = \cos\theta_1$
$S1 = \sin\theta_1$

The couple acting on O is:

$$Mg \cdot OG \cdot \cos\alpha = MgZ_G \tag{2-23}$$

and the deformation force acting on A is written as:

$$F_A = MgZ_G / L_1$$

$$F_A = g\left(M_1 \frac{1_1}{L_1} + M_2 + M_3\right) + \left(M_2 \frac{1_2}{L_1} + M_3 \frac{L_2}{L_1}\right) C1 \tag{2-24}$$

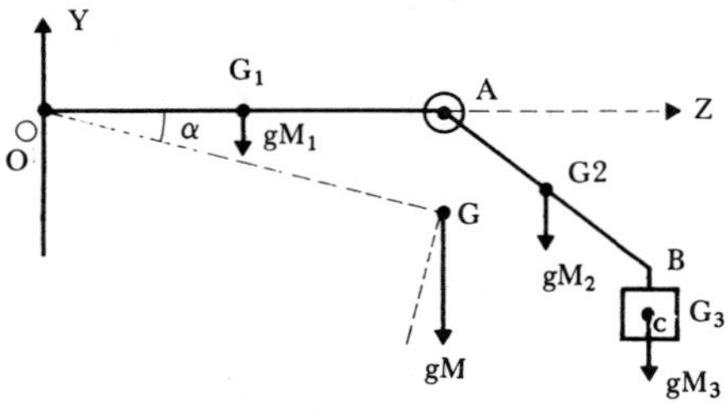

Figure 2.4.

Knowing Young's modulus and the form of OA, deformation and the movement of point A can be deduced by it. It results from the configuration ie from θ_1. A similar reasoning can be used to find the movement of B by estimating FB (A assumed fixed).

$$F_B = g\left(M_2 \frac{l_2}{L_2} + M_3\right)Cl \tag{2-25}$$

The elasticity of the articulations has been considered in Volume 1, Chapter 7. Taking it into account, even in a simplified form, seriously complicates the dynamic model affecting the vector in equation (2-2).[4] In fact, if $\underline{X}$ is the vector of elastic deformation represented by

$$\underset{N\times 1}{\underline{X}} = \underset{N\times N\ N\times 1}{R\underline{\Psi}} - \underset{N\times N\ N\times 1}{S\underline{\Theta}} \tag{2-26}$$

in which $\underline{\Psi} = (\Psi_1 \ldots \Psi_N)^T$ is the vector of the driving angles, $\underline{\Theta}$ the vector of articular variables, R and S matrices characterizing the elasticities. Then the Lagrangian of the system is written as:

$$\mathcal{L} = \mathcal{L}_\theta + \mathcal{L}_m + \underline{\Lambda}^T(\underline{X} - R\underline{\Psi} + S\underline{\Theta}) \tag{2-27}$$

where $\mathcal{L}_\theta$ is the Lagrangian of the system formed by the bodies and by the rigid articulations.
$\mathcal{L}_m$ is the Lagrangian involving the motors and the transmissions.
$\underline{\Lambda}$ is the vector of Lagrange multipliers.

It is to this new Lagrangian that Lagrange's equations must be applied:

$$\underset{i=1,N}{\frac{d}{dt}}\left(\frac{\partial \mathcal{L}}{\partial \dot{q}_i}\right) - \frac{\partial \mathcal{L}}{\partial q_i} = \Gamma_i + Q_i \tag{2-28}$$

where q_i is generalized coordinate, Γ_i is couple supplied by motor i, Q_i is couple of external forces. Note that even though equation (2-2) was obtained by applying the Lagrange method only to the part $\mathcal{L}_\theta$ of equation (2-27), it has become complex.

2.1.4 CONCLUSIONS

Even assuming that interaction of a robot with gravity can be estimated beforehand, a model describing it can only be easily derived for a rigid robot. Even then, the terms involved are non-linear and cause problems, especially when considering dynamic control. Generally, the action of gravity on elasticity is ignored, for this effect is complicated and cannot be properly observed without adding specialized sensors. This is why rigidity of the robot is a quality sought by engineers. However, for applications such as assembly work, flexibility, although undesirable for large movements (excitation of resources, precision failure), can become valuable in the assembly task proper, where there is little movement but where passive compliance allows the operation to be satisfactorily completed. Such considerations are leading to growing interest, with research being conducted into robots with *supervised rigidity*.

2.2 Interaction of robot with its medium

Unlike the case of an aeroplane moving at speed through the atmosphere, or a satellite submitted to the action of a rarified atmosphere for a long time, the effect of air resistance on terrestial robots can safely be ignored. When air is at rest and speed of movement of the robot very low, the effect of the surrounding medium is negligible. This is not the same for underwater applications of robots. Apart from problems caused by the seepage of water and corrosion of the robot's structure, the physical presence of a surrounding, incompressible fluid of considerable density is a significant factor. The following two factors must then be taken into account:

1. Effective decrease in weight, arising from Archimedes' principle.
2. (Viscous) resistance to movement in the fluid related to the density ρ and to the viscosity η of the liquid.

It is known that, during static equilibrium, pressure exerted by a fluid counteracts gravity, so that each constituent component M_i of the robot with volume V_i sees its weight effectively becoming:

$$P_i = gM_i - \rho V_i \tag{2-29}$$

The couples Q_i, produced by the force of gravity on the articulations, take the form:

$$Q_i = \sum_{\lambda=1}^{M} (gM_\lambda - \rho V_\lambda)\, f_i(\underline{\Theta}) \tag{2-30}$$

which can be compared with equation (2-3).

The forces that oppose movement in the fluid are, on the other hand, far harder to calculate, because they depend on the physical form of the moving bodies and on the flow regime of the liquid.

When a solid body moves very slowly, the fluid in which it moves flows backwards in a laminar fashion forming a kind of immobile sheath around the body. The other layers glide over this layer in such a way that the viscous forces are diminished. The resistance set up to motion is proportional to the speed of the body and the coefficient of viscosity. When the speed increases a wake is formed. The sheath is still present, but moving away from the body. The relative speed of the fluid varies very quickly, so that at a short distance from the solid body the speed becomes negligible. For this regime the force exerted by the fluid on the solid body is independent of the viscosity and is proportional to $\rho v^2/2$ (ρ = density of the fluid). For greater speeds turbulent vortices are formed, preventing simple mathematical formulation of resistance in the fluid.

2.2.1 RESISTANCE AT LOW SPEEDS

Resistance is proportional to speed, v, and viscosity, η:

$$F_R = k\eta v \quad \text{(2-31)}$$

Factor k is very difficult to calculate, for the dimensions and the form of the solid body must be taken into account. Stokes established the formula for a sphere of radius a:

$$F_R = 6\pi\eta va \quad \text{(2-32)}$$

Low speeds mean that the Reynolds number:

$$N = pvd/\eta \quad \text{(2-33)}$$

(d being a dimension of the solid body) remains low, 0.1 C.G.S. units.

In practice it is unthinkable to calculate fluid resistance for a robot. The resistance for each particular case can easily be measured experimentally. In fact this is unnecessary, for these forces are often negligible at low speeds, and can be compensated for by controls provided both at rest and at speed.

2.2.2 RESISTANCE AT GREATER SPEED

If the Reynolds number (which used to describe flow characteristics at high speeds) ($\rho = 1$, $n = 0.010$ and $N = 100$ vd in C.G.S. units for the water) is used, there are two kinds of solutions for a robot:

(a) The speeds, although high, induce resistance forces which are in fact small enough to be compensated for by kinematic control, without affecting design.
(b) The disturbances formed by the resistance of the fluid make the system of equation (2-2) collapse at higher speeds, and dynamic control is needed.

Now to complete the description of this model consider the obstacle. At speeds of interest the resistance F_R depends on:

- The geometrical parameters defining the form of the solid body.
- The constants ρ and η which characterize the fluid.
- The speed of movement v of the solid body.

Calling S the projection of the surface of the solid body on a plane normal to the speed:

$$F_R = \frac{C \cdot S\rho v^2}{2} \tag{2-34}$$

where C is a coefficient which depends on Reynolds number itself, related to the form and the direction of the body. Generally, this can only be found experimentally. Figure 2.5 gives an idea of its complexity for a simple calculable case: that of a very long cylinder of diameter d moving at speed v in a normal direction to its major axis.

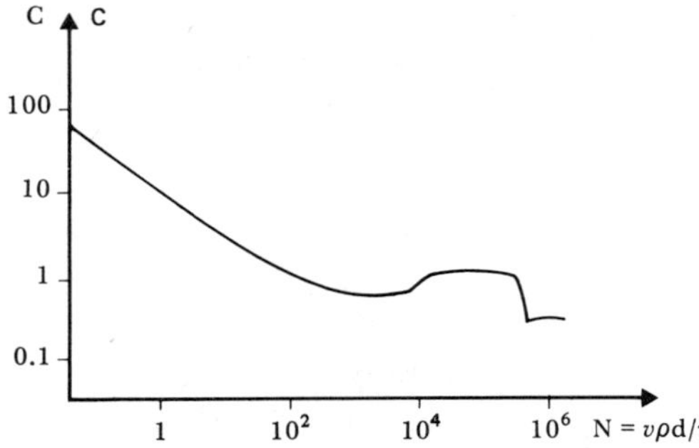

Figure 2.5. *Variation of C as a function of N for a long cylinder of diameter d moving at right angles to its axis at a speed v in a fluid characterized by* ρ *and* η[5]

Considering the various component bodies of an articulated robot, three difficulties in the derivation of F_R are encountered:

1. The bodies can have rotational movements which vary along their lengths.
2. The mutual proximity of the various solid bodies linked by the articulations makes the flow of the fluid complex – a number of simultaneous flow regimes arise.
3. The solid bodies have a finite length. This means that only measurements on the robot itself, in the fluid, will provide, in comparison with the same movements in air, correct evaluation of the resistance forces.

Therefore, when designing underwater manipulators one considers only slow-moving robots or omits any dynamic monitoring during movement, the aim being to arrive at the right place whatever the means.

2.2.3 CONCLUSIONS

The problem of fluid-robot interaction can be considered negligible in underwater use and terrestial applications. It is unlikely in future exploitations of seabed resources that these factors will be taken into account regarding design (as is done nowadays to reduce wind resistance to a minimum when designing cars). If the 'body' of the robot or its geometrical structure is to adapt automatically, this is a development that will only come about in the not-so-near future.

2.3 Interaction of robot with its load

A manipulator robot operates sometimes 'empty' ie without an object in its end effector, and sometimes 'loaded' when the gripper and the load become effectively one unit. The load represents a mass under the effect of gravity; it introduces a couple and an inertia which can create significant deviations on a dynamically pursued trajectory. It can also affect the static configuration of the robot should it not be sufficiently rigid and if the load is of significant mass (ie generating significant couples compared with others). To cope with the deviations created by the load, in the rigid robot, one can propose three principal solutions. They arise when:

1. The load can be considered to be negligible, because it is small compared with the rest of the robot.
2. The characteristics of the load and the way in which the gripper is going to handle it are known in advance. It is then possible to foresee the corrections which must be applied to take the load into account. This becomes a problem as it cannot be applied to a range of loads.

The characteristics of the load can be estimated on-line and the possible corrections of the trajectories for the presence of the load calculated. Thus load-robot interaction is fully catered for.

2.3.1 ON-LINE COMPUTATION AND THE CENTRE OF GRAVITY

The characteristics of the load that must be taken into account are:

- Mass.
- Position of the centre of gravity of that mass in relation to a set of fixed coordinates of the robot.
- Inertial tensor about the rotational axis of that mass.

If the robot control acts on its position or its speed, then the inertial tensor is not useful, for it does not contribute to the control equation. But, in this case, the mass and the position of the load's centre of gravity must still be calculated. Consider the unloaded robot, at standstill in a balanced configuration $\underline{\Theta}_{eq}$. Equation (2-2) is then written:

$$\underline{Q}(\underline{\Theta}_{eq}) + \underline{\Gamma}_{\underline{\Theta}_{eq}} = 0 \tag{2-35}$$

Using ad hoc control let the robot grasp the load, later returning into the same balance configuration $\underline{\Theta}_{eq}$. Then:

$$\underline{Q}'(\underline{\Theta}_{eq}) + \underline{\Gamma}'_{\underline{\Theta}\,eq} = 0 \tag{2-36}$$

According to equation (2-3) the vector $\underline{Q}$ has as many components as segments of the robot M. Each components Q_i takes the form:

$$Q_i = \sum_{\lambda=1}^{M} M_\lambda\, g\, \frac{\partial Y^{(0)}_{G\lambda}}{\partial \theta_i} \tag{2-3}$$

In these conditions equations (2-35) and (2-36) are written:

$$\left.\begin{aligned}
&\sum_{\lambda=1}^{M} gM_\lambda \frac{\partial Y^{(0)}_{G\lambda}}{\partial \theta_1} + \Gamma_{\theta_1} = 0 \quad \text{with } \theta_1 = \theta_{1\,eq} \\
&\quad\vdots \qquad\qquad \vdots \qquad \vdots \\
&\sum_{\lambda=1}^{M} gM_\lambda \frac{\partial Y^{(0)}_{G\lambda}}{\partial \theta_M} + \Gamma_{\theta M} = 0 \quad \text{with } \theta_M = \theta_{M_{eq}}
\end{aligned}\right\} \tag{2-37}$$

$$\left.\begin{aligned}
&\sum_{\lambda=1}^{M+1} gM_\lambda \frac{\partial Y'^{(0)}_{G\lambda}}{\partial \theta_1} + \Gamma'_{\theta_1} = 0 \quad \text{with } \theta_1 = \theta_{1_{eq}} \\
&\quad\vdots \qquad\qquad \vdots \qquad \vdots \\
&\sum_{\lambda=1}^{M+1} gM_\lambda \frac{\partial Y'^{(0)}_{G\lambda}}{\partial \theta_M} + \Gamma'_{\theta M} = 0 \quad \text{with } \theta_M = \theta_{M_{eq}}
\end{aligned}\right\} \tag{2-38}$$

In equation (2-38) M + 1 represents the load carried by the robot. As the balance configurations are identical with and without load, all the $Y'_{G\lambda}$ are equal to $Y_{G\lambda}$ except for $\lambda = M + 1$.

If equation (2-37) is subtracted from equation (2-38) then:

$$\left.\begin{aligned}
&gM_{M+1} \frac{\partial Y'^{(0)}_{G\,M+1}}{\partial \theta_1} + \Gamma'_{\theta_1} - \Gamma_{\theta_1} = 0 \\
&\quad\vdots \qquad\qquad \vdots \qquad \vdots \\
&gM_{M+1} \frac{\partial Y'^{(0)}_{G\,M+1}}{\partial \theta_M} + \Gamma'_{\theta\,M} - \Gamma_{\theta M} = 0
\end{aligned}\right\} \tag{2-39}$$

The changes in the couples exerted on the articulations are measurable.

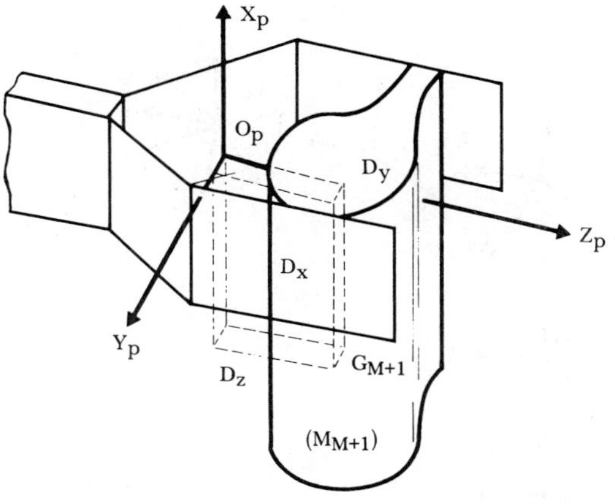

Figure 2.6. *Unsymmetrical object carried by robot gripper.* $O_pX_pY_pZ_p$ *is the coordinate trihedral of the gripper.* $D_xD_yD_z$ *are the distances of the mass load* M_{M+1} *from the centre of gravity* G_{M+1} *to the three axes*

The coordinates $Y'^{(0)}_{G_{M+1}}$ represents the height of the load's centre of gravity in the fixed coordinate R_0. This can easily be expressed in terms of the position of that load with reference to the gripper (Figure 2.6). Equation (2-39) constitutes a system of M equations resulting from four unknown variables M_{M+1}, D_x, D_y and D_z. Consideration of the couple variations allows us to find these four values. It is hypothesized here that for the robot there are as many degrees of freedom as there are limbs. In Volume 1 it was seen that a system can always be transformed to this case through the introduction of imaginary links of zero dimension and mass. Note also that equation (2-3) becomes redundant as soon as $M > 4$, which can be useful in certain specific configurations. Note furthermore that equation (2-39) is always linear for real parameter values.

Example: Consider the robot in Figure 2.7 which has six DOF. The first three bodies of mass $M_1 M_2 M_3$ form the arm which rests on the fixed body, mass M_0. As explained in Volume 1 a coordinate set as shown in the figure is associated with limbs of mass centre G_0, G_1, G_2, G_3. R_0 is the set of reference coordinates.

The end effector of mass M_6 has three DOF which represent rotations about three concurrent axes (swivel-joint). It can be represented for the calculations indicated in Figure 2.7 with imaginary masses M_4 and M_5 and zero lengths L_4 and L_5. The end effector manipulates an object of mass M_7 whose centre of gravity G_7 is fixed in the trihedral R_6 by its coordinates $D_x D_y D_z$. The value of mass M_7 and the magnitudes $D_x D_y D_z$, which are unknown, need to be determined.

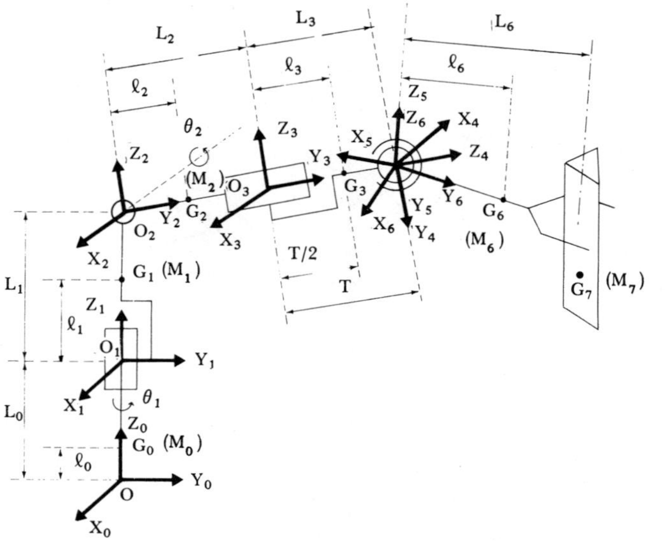

Figure 2.7a. *Robot with six DOF and associated coordinate sets*

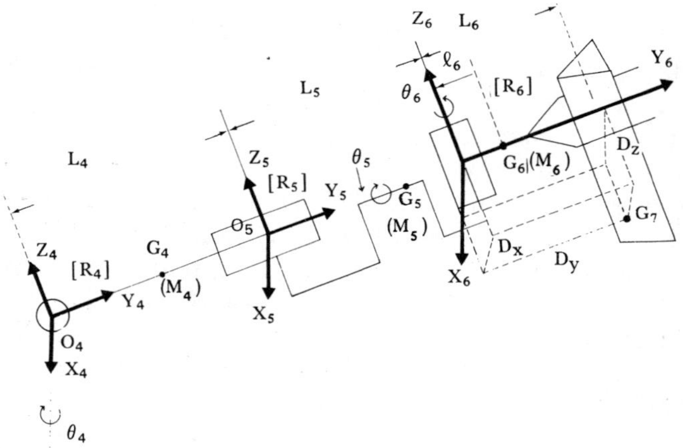

Figure 2.7b. *Representation of the end effector for the calculations:*
$L_4 = L_5 = M_4 = M_5 = 0$

The matrices, which represent the transformation from set R_{i+1} to set R_i, are successively written using the notations in section 2.1:

$$M_0^1 = \begin{pmatrix} C1 & -S1 & 0 \\ S1 & C1 & 0 \\ 0 & 0 & 1 \end{pmatrix} (\text{rotation about } Z_0) \qquad \text{(2-40)}$$

$$M_1^2 = \begin{pmatrix} 1 & 0 & 0 \\ 0 & C2 & -S2 \\ 0 & S2 & C2 \end{pmatrix} (\text{rotation about } X_1) \qquad \text{(2-41)}$$

$$M_2^3 = \begin{pmatrix} 1 & 0 & 0 \\ 0 & 1 & 0 \\ 0 & 0 & 1 \end{pmatrix} \text{(translation along } Y_2\text{)} \quad (2\text{-}42)$$

$$M_3^4 = \begin{pmatrix} 1 & 0 & 0 \\ 0 & C4 & -S4 \\ 0 & S4 & C4 \end{pmatrix} \text{(rotation about } X_3\text{)} \quad (2\text{-}43)$$

$$M_4^5 = \begin{pmatrix} C5 & 0 & S5 \\ 0 & 1 & 0 \\ -S5 & 0 & C5 \end{pmatrix} \text{(rotation about } Y_4\text{)} \quad (2\text{-}44)$$

$$M_5^6 = \begin{pmatrix} C6 & -S6 & 0 \\ S6 & C6 & 0 \\ 0 & 0 & 1 \end{pmatrix} \text{(rotation about } Z_5\text{)} \quad (2\text{-}45)$$

The coordinate of G_7 according to the vertical OZ_0 must be calculated:

$$\underline{O_6G_7}(R_0) = M_0^6\ \underline{O_6G_7}(R_6) \quad (2\text{-}46)$$

$$= M_0^1M_1^2M_2^3M_3^4M_4^5M_5^6\,[D_xD_yD_z]^T \quad (2\text{-}47)$$

Calling this coordinate $Y_{G_7}^{(0)}$ this leads to the relation:

$$\begin{aligned} Y_{G_7}^{(0)} = Y_{O_6}^{(0)} &+ D_x[C6(S2S4S5 - C2C4S5) + S6(S2C4 + C2S4)] \\ &+ D_y[C6(S2C4 + C2S4) - S6(S2S4S5 - C2C4S5)] \\ &+ D_z(C2C4C5 - S2S4C5) \end{aligned} \quad (2\text{-}48)$$

It is easy to see that:

$$Y_{O_6}^{(0)} = L_0 + L_1 + (L_2 + T)C2 \quad (2\text{-}49)$$

So:

$$\frac{\partial Y_{G_7}^{(0)}}{\partial \theta_1} = 0 \quad (2\text{-}50)$$

$$\begin{aligned} \frac{\partial Y_{G_7}^{(0)}}{\partial \theta_2} = &-(L_2 + T)S2 \\ &+ D_x[(C2S4S5 + S2C4S5)C6 + S6(C2C4 - S2S4)] \\ &+ D_y[C6(C2C4 - S2S4) - S6(C2S4S5 + S2C4S5)] \\ &+ D_z(-S2C4C5 - C2S4C5) \end{aligned} \quad (2\text{-}51)$$

$$\frac{\partial Y_{G_7}^{(0)}}{\partial \theta_3} \equiv \frac{\partial Y_{G_7}^{(0)}}{\partial T} = C2 \quad (2\text{-}52)$$

$$\frac{\partial Y^{(0)}_{G_7}}{\partial \theta_4} = D_x[C6(S2C4S5 + C2S4S5) + S6(-S2S4 + C2C4)] + D_y[C6(C2C4 - S2S4) - S6(S2C4S5 + C2S4S5)] + D_z(-C2S4C5 - S2C4C5) \quad (2\text{-}53)$$

$$\frac{\partial Y^{(0)}_{G_7}}{\partial \theta_5} = D_x[C6(S2S4C5 - C2C4C5)] - D_y S6[S2S4C5 - C2C4C5] - D_z S5(C2C4 - S2S4) \quad (2\text{-}54)$$

$$\frac{\partial Y^{(0)}_{G_7}}{\partial \theta_6} = D_x[-S6(S2S4S5 - C2C4S5) + C6(S2C4 + C2S4)] + D_y[-S6(S2C4 + C2S4) - C6(S2S4S5 - C2C4S5)] \quad (2\text{-}55)$$

Referring to equation (2-39) and adopting notations:

$$\Delta\theta_i = \Gamma_{\theta_i} - \Gamma'_{\theta_i} \quad (2\text{-}56)$$

$$a_2 = C6S5S(2+4) + S6C(2+4) \quad (2\text{-}57)$$

$$b_2 = C6C(2+4) - S6S5S(2+4) \quad (2\text{-}58)$$

$$c_2 = -C5S(2+4) \quad (2\text{-}59)$$

$$d_2 = -S2(L_2 + T) \quad (2\text{-}60)$$

$$a_5 = -C6C5C(2+4) \quad (2\text{-}61)$$

$$b_5 = S6C5C(2+4) \quad (2\text{-}62)$$

$$c_5 = -S5C(2+4) \quad (2\text{-}63)$$

$$a_6 = +S6S5C(2+4) + C6S(2+4) \quad (2\text{-}64)$$

$$b_6 = -S6S(2+4) + C6S5C(2+4) \quad (2\text{-}65)$$

One can write:

$$\Delta\Gamma_{\theta_1} = 0 \quad (2\text{-}66)$$

$$\Delta\Gamma_{\theta_2} = gM_7(a_2 D_x + b_2 D_y + c_2 D_z + d_2) \quad (2\text{-}67)$$

$$\Delta\Gamma_{\theta_3} = \Delta\Gamma_T = gM_7 C2 \quad (2\text{-}68)$$

$$\Delta\Gamma_{\theta_4} = gM_7(a_2 D_x + b_2 D_y + c_2 D_z) \quad (2\text{-}69)$$

$$\Delta\Gamma_{\theta_5} = gM_7(a_5 D_x + b_5 D_y + c_5 D_z) \quad (2\text{-}70)$$

$$\Delta\Gamma_{\theta_6} = gM_7(a_6 D_x + b_6 D_y) \quad (2\text{-}71)$$

These are five equations with four unknown variables. Equation (2-68) supplies the known mass:

$$M_7 = \Delta\Gamma_T / gC2 \tag{2-72}$$

Equations (2-69), (2-70) and (2-71) then give, for example:

$$D_x = \frac{-c_5 b_6 \Delta\Gamma_{\theta_4} + c_2 b_6 \Delta\Gamma_{\theta_5} + (b_2 c_5 - c_2 b_5)\Delta\Gamma_{\theta_6}}{gM_7 \cdot \delta} \tag{2-73}$$

$$D_y = \frac{c_5 a_6 \Delta\Gamma_{\theta_4} - c_2 a_6 \Delta\Gamma_{\theta_5} - (a_2 c_5 - a_5 c_2)\Delta\Gamma_{\theta_6}}{gM_7 \cdot \delta} \tag{2-74}$$

$$D_z = \frac{(a_5 b_6 - a_6 b_5)\Delta\Gamma_{\theta_4} - (a_2 b_6 - b_2 a_6)\Delta\Gamma_{\theta_5} + (a_2 b_5 - b_2 a_5)\Delta\Gamma_{\theta_6}}{gM_7 \cdot \delta} \tag{2-75}$$

$$\delta = a_6 (b_2 c_5 - c_2 b_5) - b_6 (a_2 c_5 - a_5 c_2) \tag{2-76}$$

From measurement of the couple variations the values of the transported mass and the position of the mass centre can be automatically found. These partially characterize the interaction of the robot with its load. Note, however, that if equations (2-68) to (2-71) are used the configurations corresponding to $\delta = 0$ or to $C2 = 0$ ie $\theta_2 = \pm\pi/2$ do not allow the calculation to be carried out.

The redundancy of the equations allows one to find M_7 from equations (2-67) and (2-69):

$$M_7 = \frac{-\Delta\Gamma_{\theta_4} + \Delta\Gamma_{\theta_2}}{(L_2 + T)S2} \tag{2-77}$$

which becomes singular for $\theta_2 = 0$ or π. In practice, one can select the configuration of the measurement beforehand, and arrange for it not to be singular or close to a singular configuration. The couples and their variations are indirectly controllable if the robot's mechanism is reversible. Consider by way of example the classical control structure in Figure 2.8.

Suppose that the robot is the one in Figure 2.7 with six DOF. $\underline{V} = (V_1 V_2 \ldots V_6)^T$ is the control vector, with for example V_1, V_2 etc being a set of voltages generated by the computer. $\underline{\epsilon} = (\epsilon_1\ \epsilon_2 \ldots \epsilon_6)^T$ is the error vector. If (K) is a diagonal matrix of 6 X 6 dimensions representing the gain of the amplifiers supplying the motors and $\underline{I} = (I_1 I_2 \ldots I_6)^T$. Suppose that the actuators are an electric motor supplying the necessary torques. Then:

$$\underline{\Gamma m} = (K')\underline{I} \tag{2-78}$$

(K′) is also a diagonal matrix 6 X 6 and $\underline{\Gamma m} = (\Gamma m_1\ \Gamma m_2 \ldots \Gamma m_6)^T$. The motors transmit their movement and their torques through reduction gears whose action is denoted by the diagonal matrix (1/N) representing position, (N) representing the couples. Nevertheless, the motors

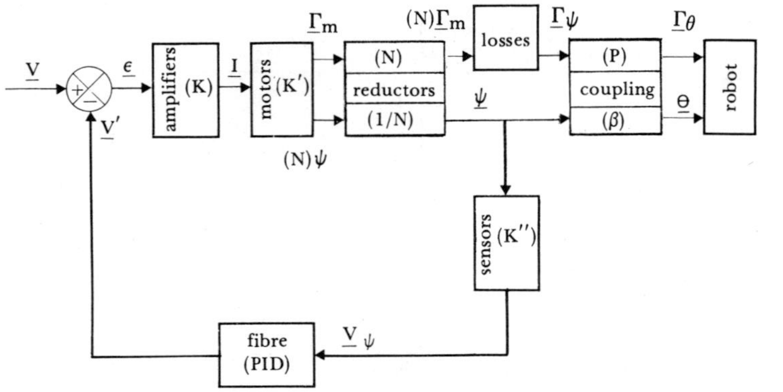

Figure 2.8. *Example of relationship between control vector $\underline{V}$ and couples with robot articulations. All values are vectors or matrices*

can have inertia represented by matrix (j) and friction similarly represented by (f). The useful couples, after gearing down, are written:

$$\underline{\Gamma}_\Psi = (N)\underline{\Gamma}m - (N)^2(j)\frac{d^2\underline{\Psi}}{dt^2} - (N)^2(f)\frac{d\underline{\Psi}}{dt} \tag{2-79}$$

After gearing down, coupling with the articular variables $\underline{\Theta}$ takes place. For example, two motors can influence the same articulation, or a variable θ_i is obtained by the sum of the action of two motors and another θ_j by their difference. Using the relation:

$$\Gamma_{\theta_i} = \sum_{j=1}^{6} \frac{\partial \Psi_j}{\partial \theta_i} \Gamma_{\Psi_i} \tag{2-80}$$

the result will be:

$$\underline{\Gamma}_\theta = (P)\,\underline{\Gamma}_\Psi \quad \text{and} \quad \underline{\Theta} = (\beta)\underline{\Psi} \tag{2-81}$$

(P) and (β) are matrices with constant coefficient.
Figure 2.8 and equations (2-78) to (2-81) provide:

$$\underline{\Gamma}_\theta = (P)(N)(K')(K)[\underline{V} - \underline{V}'] - (P)(N)^2(j)\frac{d^2\underline{\Psi}}{dt^2} - (N)^2(f)\frac{d\underline{\Psi}}{dt} \tag{2-82}$$

Suppose, moreover, that the enslavement of the motors is carried into effect by measurement of and correction of type PID. Then:

$$\underline{V}_\Psi = (K'')\underline{\Psi} \tag{2-83}$$

(K'') being a diagonal matrix with constant coefficient and:

$$\underline{V}' = (K_1)\underline{V}_\Psi + (K_2)\frac{d\underline{V}_\Psi}{dt} + (K_3)\int \underline{V}_\Psi \, dt \qquad (2\text{-}84)$$

The combination of equations (2-82) and (2-84) gives a relation between the couple vector $\underline{\Gamma}_\theta$, the control vector $\underline{V}$ and the angle vector $\underline{\Psi}$. Suppose:

$$\begin{aligned} \underline{\Gamma}_\theta = {} & (P)(N)(K')(K)\underline{V} - (P)(N)(K')(K)(K_1)(K'')\underline{\Psi} \\ & - [(P)(N)(K')(K)(K_2)(K'') + (P)(N)^2(f)]\frac{d\underline{\Psi}}{dt} \\ & - (P)(N)^2(j)\frac{d^2\underline{\Psi}}{dt^2} - (P)(N)(K')(K)(K_3)(K'')\int \underline{\Psi}\, dt \end{aligned} \qquad (2\text{-}85)$$

In measuring the parameters of the transported mass, difficulties will arise; first, when the end effector is stopped when empty, then when loaded, in the same configuration. The difference between the two values of equation (2-85) is therefore expressed as:

$$\underline{\Delta\Gamma}_\theta = (P)(N)(K')(K)\underline{\Delta V} \qquad (2\text{-}86)$$

Returning to equation (2-72):

$$M_7 = \Delta\Gamma_{\theta_3}/gC2$$

Supposing that (P) is not diagonal, the mass will be determined using:

$$\begin{pmatrix} \Delta\Gamma_{\theta_1} \\ \Delta\Gamma_{\theta_2} \\ \cdot \\ \cdot \\ \cdot \\ \Delta\Gamma_{\theta_6} \end{pmatrix} = \begin{pmatrix} P_{11}P_{12} \ldots P_{16} \\ P_{21}P_{22} \ldots \\ \cdot \\ \cdot \\ \cdot \\ P_{61}P_{62} \ldots P_{66} \end{pmatrix} \begin{pmatrix} n_1 k'_1 k_1 & & & & \\ & n_2 k'_2 k_2 & & & \\ & & \cdot & & \\ & & & \cdot & \\ & & & & \cdot \\ & & & & & n_6 k'_6 k_6 \end{pmatrix} \begin{pmatrix} \Delta V_1 \\ \Delta V_2 \\ \cdot \\ \cdot \\ \cdot \\ \Delta V_6 \end{pmatrix} \qquad (2\text{-}87)$$

and supposing that:

$$M_7 = \frac{1}{gC2}\sum_{i=1}^{6} P_{3i} n_i k'_i k_i \, \Delta V_i \qquad (2\text{-}88)$$

the same procedure applies to the calculation of $D_x D_y D_z$. The relationship between the unknown parameters of the transported load and the easily accessible magnitudes ΔV_i have been well established. The more elastic the transmissions and the greater the mass, the greater the error.

2.3.2 ON-LINE COMPUTATION OF THE INERTIAL TENSOR OF THE LOAD

In the case of a perfectly rigid robot when dynamic control is required (as discussed in Volume 1, there are many cases that cannot be con-

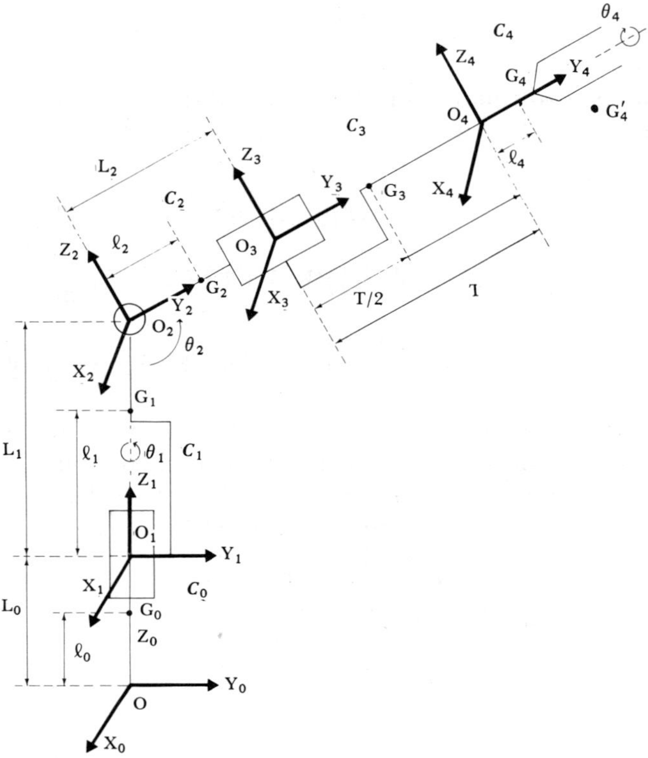

Figure 2.9. *Robot with four DOF. M_1, M_2, M_3, M_4 are the masses of the four mobile bodies. G_4 is the empty mass centre of the gripper when empty. G'_4 is the loaded mass with its centre fixed in the trihedral $O_4X_4Y_4Z_4$ by the coordinates $D_xD_yD_z$. The inertial tensor of each body i about its three rotational axes is written:*

$$\underline{\underline{I}}^{(K)} = \begin{vmatrix} I_{11}^{(K)} & I_{12}^{(K)} & I_{13}^{(K)} \\ I_{21}^{(K)} & I_{22}^{(K)} & I_{23}^{(K)} \\ I_{31}^{(K)} & I_{32}^{(K)} & I_{33}^{(K)} \end{vmatrix}$$

trolled) a procedure similar to that of section 2.3.1 may be used to determine the inertial tensor of the load. By way of example, consider the manipulator in Figure 2.9 which has four DOF (for simplicity's sake). The end effector only has a rotational movement θ_4 about Y_4. In Volume 1, Chapter 7 the derivation of description of the dynamic model for such a system was indicated. The calculation shows that the system of equations here has the form:

$$\begin{pmatrix} a_{11} & a_{12} & a_{13} & a_{14} \\ a_{12} & a_{22} & a_{23} & a_{24} \\ a_{13} & a_{23} & a_{33} & a_{34} \\ a_{14} & a_{24} & a_{34} & a_{44} \end{pmatrix} \begin{pmatrix} \ddot{\theta}_1 \\ \ddot{\theta}_2 \\ \ddot{T} \\ \ddot{\theta}_4 \end{pmatrix} + \begin{pmatrix} 0 & c_{12} & 0 & c_{14} \\ c_{21} & 0 & c_{23} & c_{24} \\ c_{31} & c_{32} & 0 & c_{34} \\ 0 & c_{42} & 0 & 0 \end{pmatrix} \begin{pmatrix} \dot{\theta}_1^2 \\ \dot{\theta}_2^2 \\ \dot{T}^2 \\ \dot{\theta}_4^2 \end{pmatrix} +$$

$$\begin{pmatrix} -2c_{21} & -2c_{31} & 0 & b_{14} & b_{15} & b_{16} \\ 0 & b_{22} & b_{23} & b_{24} & b_{25} & b_{26} \\ -b_{22} & 0 & b_{33} & 0 & -b_{26} & 0 \\ -b_{23} & -b_{33} & 0 & b_{26} & 0 & b_{46} \end{pmatrix} \begin{pmatrix} \dot\theta_1\dot\theta_2 \\ \dot\theta_1\dot{T} \\ \dot\theta_1\dot\theta_4 \\ \dot\theta_2\dot{T} \\ \dot\theta_2\dot\theta_4 \\ \dot{T}\dot\theta_4 \end{pmatrix} = \begin{pmatrix} Q_1 + \Gamma\theta_1 \\ Q_2 + \Gamma\theta_2 \\ Q_3 + \Gamma\theta_3 \\ Q_4 + \Gamma\theta_4 \end{pmatrix} \tag{2-89}$$

If a single articulation only is working at one time, the manipulator's movement will be described by only one of the following four equations, if θ_2, T and θ_4 are constant:

$$a_{11}\ddot\theta_1 = Q_1 + \Gamma_{\theta_1} \tag{2-90}$$

If θ_1, T and θ_4 are constant:

$$a_{22}\ddot\theta_2 = Q_2 + \Gamma_{\theta_2} \tag{2-91}$$

Likewise by mobilizing T only:

$$a_{33}\ddot{T} = Q_3 + \Gamma_T \tag{2-92}$$

and moving θ_4 only:

$$a_{44}\ddot\theta_4 = Q_4 + \Gamma_{\theta_4} \tag{2-93}$$

Now write the expressions for the dynamic coefficients a_{ii} and the gravitational couples:

$$\begin{aligned} a_{11} = {} & 2M_4\,[(T^2/4)(1 - 2S2C2) + D_x^2(C^2 4C^2 2 + S^2 2) + D_y^2 C^2 2 \\ & + D_z^2(S^2 4C^2 2 + S^2 2) + 2D_xD_y\, S2C2S4 - 2D_yD_z\, S2C2C4 \\ & + 2D_xD_z\, C^2 2S4C4 + TD_xS2(C2 - S2)S4 + TD_yC2(C2 - S2) \\ & + TD_zS2(C2 - S2)C4] + 2[I_{11}^{(4)}S^2 4C^2 2 - I_{21}^{(4)}S4C2S2 \\ & - I_{31}^{(4)}S4C4C^2 2 - I_{12}^{(4)}S2C2S4 + I_{22}^{(4)}S^2 2 + I_{32}^{(4)}C2S2C4 \\ & - I_{13}^{(4)}S4C4C^2 2 + I_{23}^{(4)}C2S2C4 + I_{33}^{(4)}C2C4 + I_{33}^{(1)} + I_{22}^{(2)}S^2 2 \\ & + I_{33}^{(2)}C^2 2 + I_{22}^{(3)}S^2 2 + M_2 1_2^2\, S^2 2 + M_3\, S^2 2\, T^2/4] \end{aligned} \tag{2-94}$$

$$\begin{aligned} a_{22} = {} & 2M_4\,[(T^2/2) + D_x^2\, S^2 4 + D_y^2 + D_z^2\, C^2 4 - TD_xS4 + TD_y \\ & + TD_zC4 - 2D_xD_zS4C4] + 2[I_{11}^{(4)}C^2 4 + I_{33}^{(4)}S^2 4 + I_{13}^{(4)}S4C4 \\ & + I_{31}^{(4)}S4C4 + I_{11}^{(2)} + I_{11}^{(3)} + M_2 1_2^2 + M_3T^2/4] \end{aligned} \tag{2-95}$$

$$a_{33} = 2(M_3 + M_4) \tag{2-96}$$

$$a_{44} = 2M_4(D_x^2 + D_z^2 + I_{22}^{(4)}) \tag{2-97}$$

$$Q_1 = gM_4\,[D_x(C1C4 + S1S2S4) - D_yC1C2 - D_z(C1S4 - S1S2C4)] \tag{2-98}$$

$$Q_2 = -gM_2 l_2 S2 - gM_3 S2(L_2 + T/2) - gM_4 [(L_2 + T)S2 + D_x C1C2S4 + D_y C1C2 + D_z C1C2C4] \quad (2\text{-}99)$$

$$Q_3 = gM_4 C2 + gM_3 C2/2 \quad (2\text{-}100)$$

$$Q_4 = -gM_4 [D_x (S1S4 + C1S2C4) + D_z (S1C4 - C1S2S4)] \quad (2\text{-}101)$$

One sees that the coefficients a_{ii} (except for a_{33}) result from the elements of the inertial tensor of the end effector. But only a_{ii} includes the nine components. Choose therefore to move only by influencing θ_1. Equation (2-94) has the form (θ_2, θ_4 and T being condensed to a known value which does not annul any of the terms involved):

$$a_{11} = M_4 (k_1 + k_2 D_x^2 + k_3 D_y^2 + k_4 D_z^2 + k_5 D_x D_y + k_6 D_y D_z + k_7 D_x D_z + k_8 D_x + k_9 D_y + k_{10} D_z) + k_{11} I_{11}^{(4)} + k_{12} I_{21}^{(4)} + k_{13} I_{31}^{(4)} + k_{14} I_{12}^{(4)} + k_{16} I_{22}^{(4)} + k_{17} I_{32}^{(4)} + k_{18} I_{13}^{(4)} + k_{19} I_{23}^{(4)} + k_{20} I_{33}^{(4)} + k_{21} \quad (2\text{-}102)$$

To examine the determination procedure of $\Gamma_{ij}^{(4)}$ bring the empty robot into a configuration C1 and apply couple $\Delta\Gamma\theta_1$. The robot will move and then stop in a configuration C2. During this movement care will be taken to record $\theta_1(t)$. Applying it to configuration C1, this time with the load in position, the same couple $\Delta\Gamma\theta_1$ is applied. Now record $\theta_1(t)$ again during the movement. The comparison of the two recordings allows the inertial tensor of the load to be found again. In the first case: M_4 is the mass of the gripper ($D_x = D_2 = 0$; $D_y = I_4$). The $I_{ij}^{(4)}$ are those of the gripper and the equation describing the response to the couple is:

$$[M_4 (k_1 + k_3 l_4^2 + k_9 l_4) + k_{11} I_{11}^{(4)} + k_{12} I_{21}^{(4)} + k_{13} I_{31}^{(4)} + k_{14} I_{12}^{(4)} + k_{16} I_{22}^{(4)} + k_{17} I_{32}^{(4)} + k_{18} I_{13}^{(4)} + k_{19} I_{23}^{(4)} + k_{20} I_{33}^{(4)} + k_{21}]\, \ddot{\theta}_1 = gM_4 l_4 m_3 C1 + \Gamma_{\theta_1} \text{ (with } m_3 = -C2) \quad (2\text{-}103)$$

In the second case call M_4^0 the mass of the inertial load $I_{ij}'^{(4)}$. The equation of movement is written:

$$[(M_4 + M_4')(k_1 + k_2 D_x^2 + k_3 D_y^2 + k_4 D_z^2 + k_5 D_x D_y + k_6 D_y D_z + k_7 D_x D_z + k_8 D_x + k_9 D_y + k_{10} D_z) + k_{11} (I_{11}^{(4)} + I_{11}'^{(4)}) + k_{12} (I_{21}^{(4)} + I_{21}'^{(4)}) + k_{13} (I_{31}^{(4)} + I_{31}'^{(4)}) + k_{14} (I_{12}^{(4)} + I_{12}'^{(4)}) + k_{16} (I_{22}^{(4)} + I_{22}'^{(4)}) + k_{17} (I_{32}^{(4)} + I_{32}'^{(4)}) + k_{18} (I_{13}^{(4)} + I_{13}'^{(4)}) + k_{19} (I_{23}^{(4)} + I_{23}'^{(4)}) + k_{20} (I_{33}^{(4)} + I_{33}'^{(4)}) + k_{21}]\, \ddot{\theta}_1 = g(M_4 + M_4')[D_x (m_1 C1 + m_2 S1) + D_y m_3 C1 + D_z (m_4 C1 + m_5 S1)] + \Gamma_{\theta_1} \quad (2\text{-}104)$$

Provided the determination procedure above has been applied, M_4', D_x,

D_y, D_z, equations (2-103) and (2-104) take the form:

$$k'_1 \ddot{\theta}_1 = m'_1 \, C1 + \Gamma_{\theta_1} \qquad \textbf{(2-105)}$$

k'_1, m'_1 and Γ_{θ_1} are then known constants:

$$[k'_2 + k_{11} I'^{(4)}_{11} + k_{12} I'^{(4)}_{21} + k_{13} I'^{(4)}_{31} + k_{14} I'^{(4)}_{12} + k_{16} I'^{(4)}_{22} + k_{17} I'^{(4)}_{32} + k_{18} I'^{(4)}_{13} + k_{19} I'^{(4)}_{23} + k_{20} I'^{(4)}_{33}] \, \ddot{\theta}_1 = m'_2 C1 + m'_3 S1 + \Gamma_{\theta_1} \qquad \textbf{(2-106)}$$

All the constants k'_2, k_{ij}, m'_i and Γ_{θ_1} are known. If the development of θ_1 at the beginning, for condition $\Gamma_{\theta_1} = \Gamma_1$, is recorded, for every moment comprising that response, the following is obtained from equation (2-106):

$$k_{11} I'^{(4)}_{11} + k_{12} I'^{(4)}_{21} + k_{13} I'^{(4)}_{31} + k_{14} I'^{(4)}_{12} + k_{16} I'^{(4)}_{22} + k_{17} I'^{(4)}_{32} + k_{18} I'^{(4)}_{13} + k_{19} I'^{(4)}_{23} + k_{20} I'^{(4)}_{33} = \frac{m'_2 \cos(\theta_1(t)) + m'_3 \sin(\theta_1(t)) + \Gamma_1 - k'_2}{\ddot{\theta}_1(t)} \qquad \textbf{(2-107)}$$

Thus all that needs to be done is to find θ_1 at various different instants, and to calculate the value of the second derivative at these instants (Figure 2.10). In equation (2-107) the right-hand term has a certain value A^1.

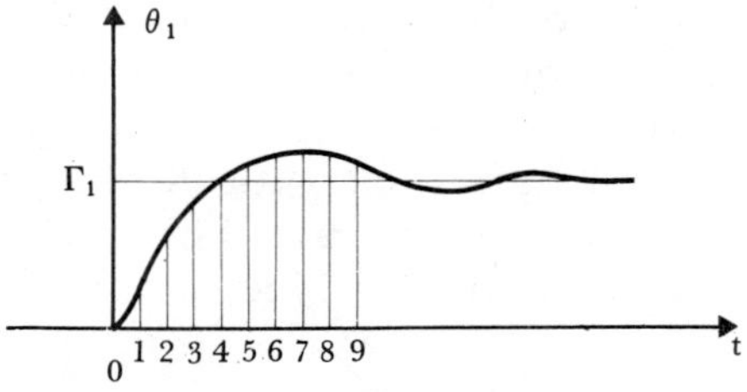

Figure 2.10. *Response of the system to a step in couple* Γ_1

If the terms k_{11} to k_{20}, m'_2, m'_3 and k'_2 are seen to depend on θ_{21} and θ_4, eight combinations of these variables can be chosen, from which eight equations of the same type as equation (2-107) can be formed.

Thus there will be nine linear equations with nine unknowns of the following type:

$$k^j_{11} \, I'^{(4)}_{11} + \ldots \, k^j_{20} \, I'^{(4)}_{33} = A^j$$

$$j = 1 \text{ to } 9$$

The $I'^{(4)}_{ij}$ are found by a simple matrix inversion.

It should be remembered that this method is valid for a given gripping position only and the process should be repeated if grasping ‘from a different angle’. Furthermore, the method is based on a ‘rigid robot’ model and begins to collapse (unless one models the elasticities perfectly) as soon as the robot has some flexibility in its limbs.

Conclusions

In this chapter the reciprocal effects that the robot and gravity and surrounding media exert on each other have been described. In the case of gravity it is obvious that the effect of the robot’s own mass is completely negligible and that it is more a matter of the action of gravity on the robot than of an interaction.

The gravitational effect of the robot’s structure in applications where gravity is zero, as in space, should however be considered. Then, the movements of the robot can create forces equivalent to an artificial gravity.

Fluids, on the other hand, have a more significant effect on the action of the robot, at least fluids whose density approaches that of water. We have seen that the dynamic control of a robot would be extremely complicated in such circumstances because modelling the reaction of the fluid is very difficult, indeed impossible.

Finally, the interaction of the robot with the transported load is usually significant, and we have seen that static or even dynamic elements of this load may quite easily be taken into account automatically by the robot itself. This is a point of some significance. Having clarified this point, the way is now clear for showing in ensuing chapters that touch can be artifically reconstructed in order to control contact between the environment and the machine.

Chapter 3

Interactions involving physical contact between robot and environment: the detection of stress

When an articulated robot is in physical contact with a solid object, the conditions necessary for an appropriate reaction from the robot are:

1. It must be able to recognize that there actually is contact (*ie detection*).
2. It must know where the points of contact are (*ie localization*).
3. It must know the characteristics of the contacts eg be able to assess the forces that have developed (*ie characterization*).

When these three factors are known (all are relevant to the final task target) the robot will be able to compute or apply a behaviour strategy which allows it to progress towards the aim assigned to it.

Two categories of physical contact are recognized:

(a) Those which act on one or several points and which are principally going to serve as detectors for safety, position measurement or as transducers for stress detection.

(b) Contacts that actuate many contact points simultaneously (eg a tactile matrix sensor) for environment assessment and recognition of the environment.

This is why physical contact sensors in a robot are generally classified either (i) as *tactile sensors* (emphasizing the objective of recognizing the environment, especially its geometric properties) or (ii) as *stress sensors.* Although stress sensors are often used as security devices (couple restrictors) their primary duty is to assess stress to guide robot activity. Stress sensors are also commonly employed in component assembly processes, where tolerances beyond the accuracy of positional control need to be achieved.

In this chapter the detection and control of stress by a robot will be discussed. Chapter 4 will deal with tactile recognition of the environment.

3.1 Preliminary hypothesis concerning detection of stress

When two objects are in contact the contact is never at a single point.

In what follows it will be supposed that there is an area of contact between robot and object and that this can be approximated by point-contact considerations. In fact, as soon as there are several zones of contact the reaction forces become difficult to assess for each zone and one can only use global data.

The factors that are going to be of interest are those which allow the resultant of reaction forces between objects to be calculated. The point of application, the amplitude and the direction of this resultant reaction force have to be known. It has been seen that all calculations involved in the control of robots, in the last analysis, are referred to a reference coordinate set R_0 related to the robot (Figure 3.1). Interaction is characterized by six variables: $X_0(P)$, $Y_0(P)$, $Z_0(P)$, F_{x0}, F_{y0}, F_{z0}. To evaluate the six variables one needs to employ — (i) a sensor that identifies the position of P in R_0 and (ii) a three-dimensional sensor that identifies the three components of force F with reference to the axes of R_0. The identification of these six variables can be effected in various ways, depending on the problem to be solved.

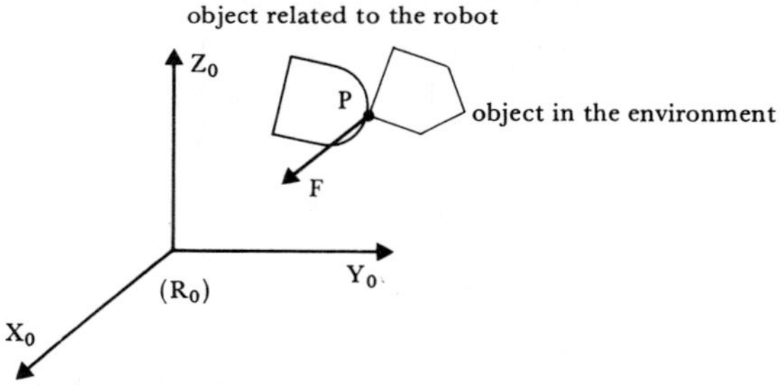

Figure 3.1. *Reaction F to be identified in coordinate set R_0*

3.2 Determination of location of the point of interaction knowing force F and its moments with reference to the axes of R_0

Consider Figure 3.2 in which force $\underline{F}$ is resolved into its three components F_{x0}, F_{y0}, F_{z0} and for which its moments are taken with respect to the three axes. The three moments are written:

$$M_{x0} = F_{z0}Y_0 - F_{y0}Z_0 \tag{3-1}$$

$$M_{y0} = F_{x0}Z_0 - F_{z0}X_0 \tag{3-2}$$

$$M_{z0} = F_{y0}X_0 - F_{x0}Y_0 \tag{3-3}$$

Going back to the coordinates of point P (to which $\underline{F}$ is applied) it can be seen that the three equations are redundant and form only a second-order system:

$$F_{x0}M_{x0} + F_{y0}M_{y0} + F_{z0}M_{z0} = 0 \tag{3-4}$$

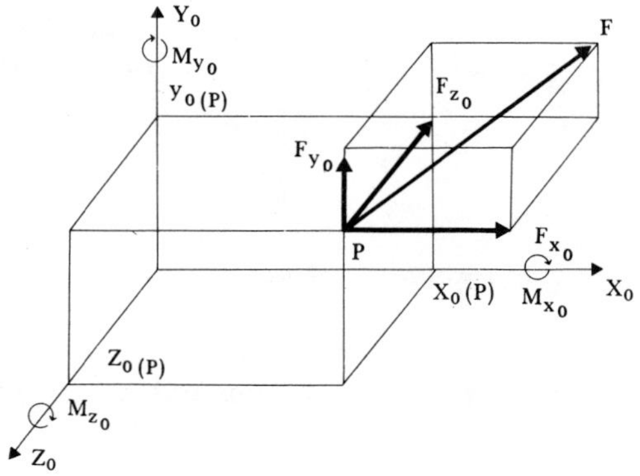

Figure 3.2. *Forces and moments*

So only two coordinates can be found. This is often sufficient (eg in inertial platforms, where one looks for the projection of the centre of gravity), the height being known. So:

$$X_0(P) = [F_{x0} Y_0(P) - M_{z0}]/F_{y0} \tag{3-5}$$

$$Z_0(P) = [F_{z0} Y_0(P) + M_{x0}]/F_{y0} \tag{3-6}$$

3.3 Interaction strategies

In all cases robot-environment interaction is expressed as a force-position relationship ie a positional increment $\underline{\Delta X}$ of point P (on the manipulator or the solid object carried by the end effector) translates into an increment $\underline{\Delta F}$ in the reaction force of the environment.

When using a robot the two main aims are:

1. To attain a *specific position* (as in an assembly task). The movements produced as a result of controlled efforts have as aim the reduction to a minimum of the reaction force (or its normal components), opposing the movement.
2. To attain a *specific force* (eg in buffing or machine-finishing). Movements of the robot depend on the difference between this force and the actual measured force.

In these cases one can speak of *piloting* by stress, which is effectively the same as *piloting* by the environment. It must, however, be remembered that the relationship $\underline{\Delta F} - \underline{\Delta X}$ applies to variations $\underline{\Delta F}$ and $\underline{\Delta X}$ applied to a known point P only, so if this relationship is to be applied its location must be approximated.

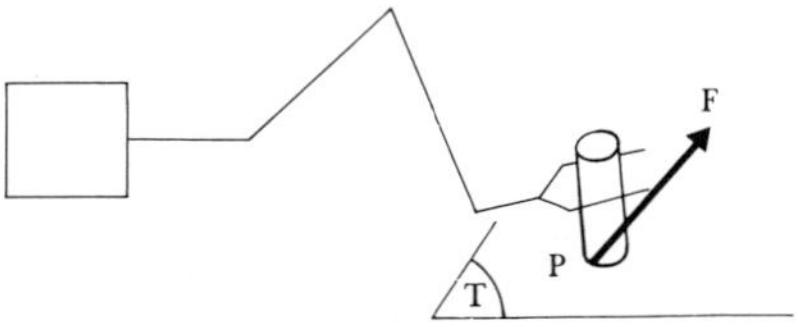

Figure 3.3. *Reaction force between table and object*

3.4 Principles of stress detection between articulated robot and environment

Consider a robot which holds an object in its end effector in contact with the environment (eg the work plane in Figure 3.3). To find the force exerted between plane T and object M, there are three possible procedures:

1. *Equip the environment* with sensors. The object could then be placed on a platform fitted with such detectors (Figure 3.4).
2. Place instrumentation on the wrist of the robot (Figure 3.6).
3. *Use the actuators as sensors*, which may be done if the robot is reversible (then the feedback from stress applied to the end effector acts on its actuators through the servos).

These three procedures will now be examined in more detail.

3.4.1. THE USE OF AN INSTRUMENTATED PLATFORM

The force exerted between robot and environment can be measured using a platform equipped with sensors. The platform in Figure 3.4 uses three plates. Between the central plate and the top plate there are sen-

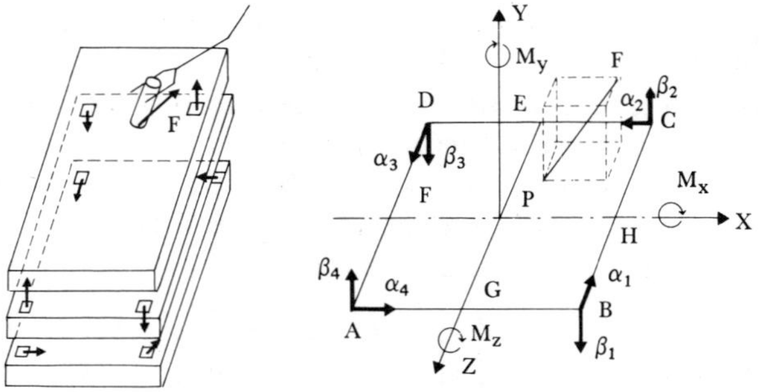

Figure 3.4. *Example of instrumentated platform for detecting stress between objects and platform*

sors that allow the vertical forces to be measured. Between the central plate and the bottom plate there are sensors that allow horizontal forces to be measured.

If the measured horizontal forces are called α_i and the vertical forces called β_i, one can see from Figure 3.4 that:

$$F_x = \alpha_4 - \alpha_2 \quad (3\text{-}7)$$

$$F_y = \beta_2 + \beta_4 - \beta_1 - \beta_3 \quad (3\text{-}8)$$

$$F_z = \alpha_3 - \alpha_1 \quad (3\text{-}9)$$

$$M_x = [(\beta_4 + \beta_3) - (\beta_1 + \beta_2)]\ell \quad (3\text{-}10)$$

$$M_y = [\alpha_1 + \alpha_2 + \alpha_3 + \alpha_4]\ell \quad (3\text{-}11)$$

$$M_z = [(\beta_4 + \beta_1) - (\beta_3 + \beta_2)]\ell \quad (3\text{-}12)$$

Supposing that the sensors are distributed over a square, side 2ℓ. The point of application of force F will then have coordinates:

$$Y_p = \theta; \qquad X_p = -M_z/F_y; \qquad Z_p = M_x/F_y \quad (3\text{-}13)$$

A platform of this kind has been used by Watson and Drake[6] to carry out assembly work.

Suppose a cylinder has to be inserted with low tolerance into a bore. The platform carries the recipient part ie the bore that has a known geometry. Its direction is known but not its position (Figure 3.5). The sensors monitor six components of magnitudes F_x, F_y, F_z, M_x, M_y, M_z from which the position of the centre of gravity may be found:

$$F_x = F_z = 0 \quad (3\text{-}14)$$

$$F_y = \beta_2 + \beta_4 - \beta_1 - \beta_3 \quad (3\text{-}15)$$

$$M_y = 0 \quad (3\text{-}16)$$

$$M_x = [(\beta_4 + \beta_3) - (\beta_1 + \beta_2)]\ell \quad (3\text{-}17)$$

$$M_z = [(\beta_4 + \beta_1) - (\beta_2 + \beta_3)]\ell \quad (3\text{-}18)$$

from which is derived:

$$X_G = \ell[\beta_4 + \beta_1 - \beta_2 - \beta_3]/[\beta_4 - \beta_1 + \beta_2 - \beta_3] \quad (3\text{-}20)$$

$$Z_G = \ell[\beta_4 + \beta_3 - \beta_1 - \beta_2]/[\beta_4 - \beta_3 - \beta_1 + \beta_2] \quad (3\text{-}19)$$

Using another process the robot must now grip the cylinder and bring it to the correct height close to the insertion position. But first, positioning movements need to be made until the entrance to the bore is located, followed by downward locating movements which maintain neutrality of forces F_x and F_z detected by the platform.

Successful insertion results when the robot reaches the appropriate final configuration or if F_y reaches a desired threshold conserving $F_x = F_z = 0$ without further movement.

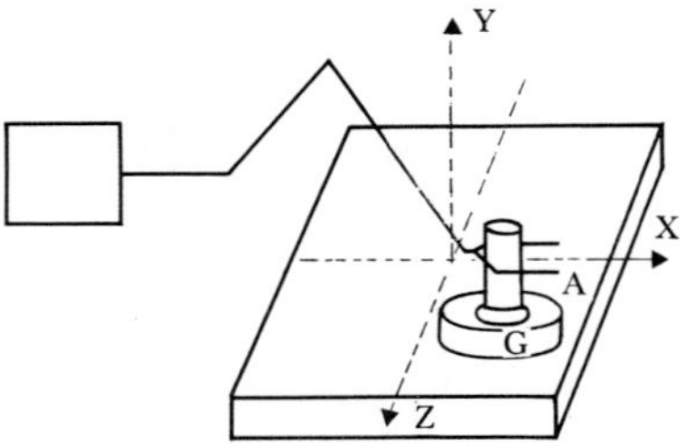

Figure 3.5. *Determination of the position of a piece A placed on an instrumentated platform*

3.4.2 USE OF A DETECTOR FITTED TO THE ROBOT

Instrumentating the environment is always difficult when precision is involved. One would always prefer to buy a complete machine or robot without having to fit delicate sensors to the task area. This is why the robot is often equipped with sensors that can fulfil the same, or nearly the same, functions as platform sensors. The additional sensors are best located just behind the end effector, at the wrist of the robot ie as near as possible to the region where stress is exerted (Figure 3.6).

The aim is to measure the couples and the forces, with reference to the three standard axes, of the end. An instrumentated wrist consists of a structure comprising compliant components and sensors that measure the deformation of these compliant parts with reference to three orthogonal axes, when one force or one couple is applied.

Sensors can be of different kinds: strain gauges, piezoelectric, magnetostrictive, magnetic and other transducers.[7] Strain gauges are reliable, cheap and easy to employ and have been used in numerous applications (Watson and Drake,[6] Goto *et al.*,[8] Rosen and Nitzan,[9] Groome,[10] Hill and Sword,[11] Bejczy[12]).

The principle is identical in each case, but each detector has its own variant which is chosen either for precision of the main measurement (encoded into 4 to 12 bits according to the model) or for its practicability. In fact, these detectors are designed principally for use in the assembly of high tolerance parts. To carry out work of this kind success-

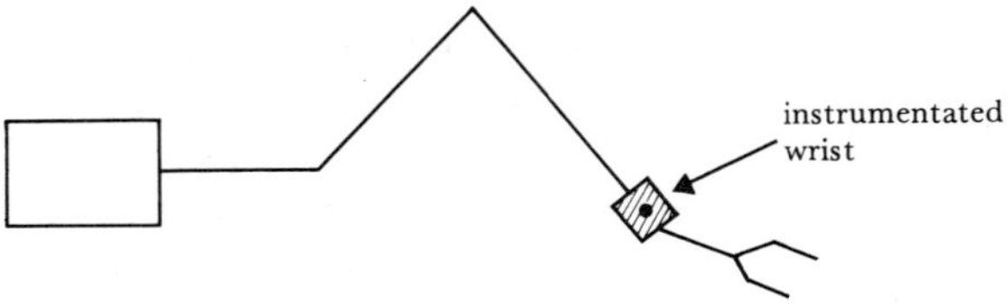

Figure 3.6. *Measurement of stress by instrumentated robot*

fully, there are various strategies: those arising from the robot itself, from the assembly task, from constraints arising from the nature of the environment etc.

By way of example, the work of Goto illustrates some of these factors. He was able to drive, in a vertical direction, a cylinder of 13-mm diameter into a bore with a low diameter tolerance (of 20 μm). He used not only *active* compliance of the detector (ie control of micromovements using measurements of stress), but also *passive* compliance (ie natural movements of the end effector caused by the stress exerted on it and its own flexibility). The reaction forces directed the cylinder towards the final required position.

To illustrate active and passive compliances and their use, refer to Figure 3.7 where it is supposed, for simplicity, that everything happens according to plan.

In active compliance (c), the whole robot moves to try to neutralize stresses on the sensors. This sets the ferrule in the right direction for insertion. Use of passive compliance (d) employs elastic deformation of the detector in the assembly process despite a direction mismatch of the ferrule with respect to the bore.

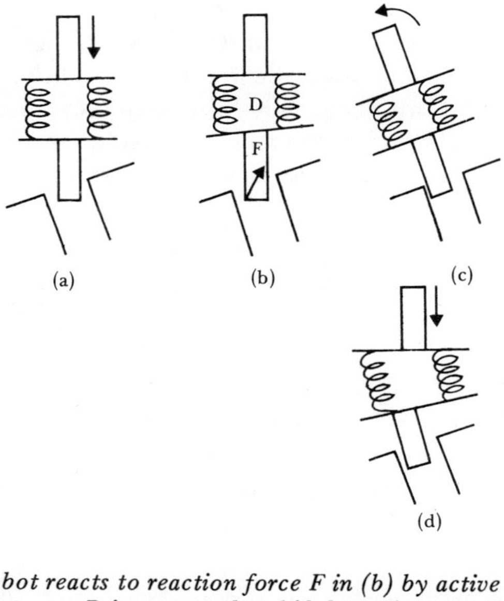

Figure 3.7. *The robot reacts to reaction force F in (b) by active compliance if the deformation of detector D is measured and if the active assembly is turned until this deformation is neutralized (c). It will react, with the help of passive compliance, if the detector can be sufficiently deformed to exert vertical stress on it. The assembly is then completed without damage (d). In this case there is no need to measure deformation of the detector except for safety reasons*

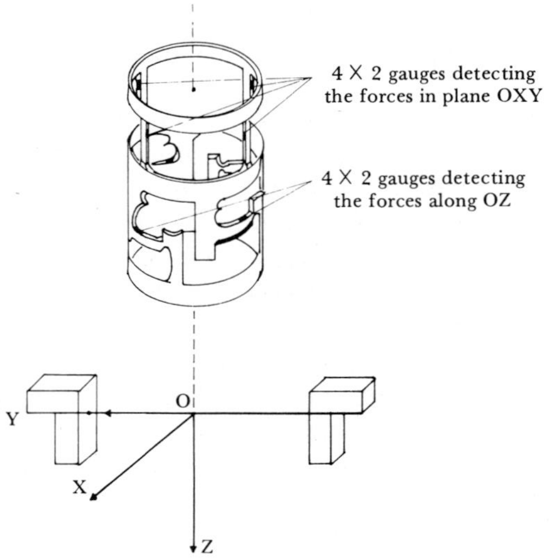

Figure 3.8. *Stress sensors experimented with at Stanford Research Institute (wrist force sensor). The detector is integral with the gripper*

Detectors of a passive compliance type may be found which are designed for insertion tasks.[13]

As an example, here we look at some active compliance systems assembled by Rosen and Nitzan[9] and Bejczy[12] (Figures 3.8 and 3.9). The sensor used by Rosen is a hollow and cut cylinder of the kind that could be used with bridge gauges which detect forces resolved on three reference axes. Bejczy proposes the system shown in Figure 3.9. It is in the form of a cross whose arms contain sensors and is integral with a cylinder.

If $W_1 \ldots W_8$ are the measured signals derived from the different sensors, then the following expression for the three reference axes of the sensor is derived:

$$\begin{pmatrix} F_x \\ F_y \\ F_z \\ M_x \\ M_y \\ M_z \end{pmatrix} = \begin{pmatrix} 0 & 0 & K_{13} & 0 & 0 & 0 & K_{17} & 0 \\ K_{21} & 0 & 0 & 0 & 0 & K_{25} & 0 & 0 \\ 0 & K_{32} & 0 & K_{34} & 0 & K_{36} & 0 & K_{38} \\ 0 & 0 & 0 & K_{44} & 0 & 0 & 0 & K_{48} \\ 0 & K_{52} & 0 & 0 & 0 & K_{56} & 0 & 0 \\ K_{61} & 0 & K_{63} & 0 & K_{65} & 0 & K_{67} & 0 \end{pmatrix} \begin{pmatrix} W_1 \\ W_2 \\ W_3 \\ W_4 \\ W_5 \\ W_6 \\ W_7 \\ W_8 \end{pmatrix} \quad (3\text{-}21)$$

where K_{ij} are constant characterizing gains of the various bridge gauges.

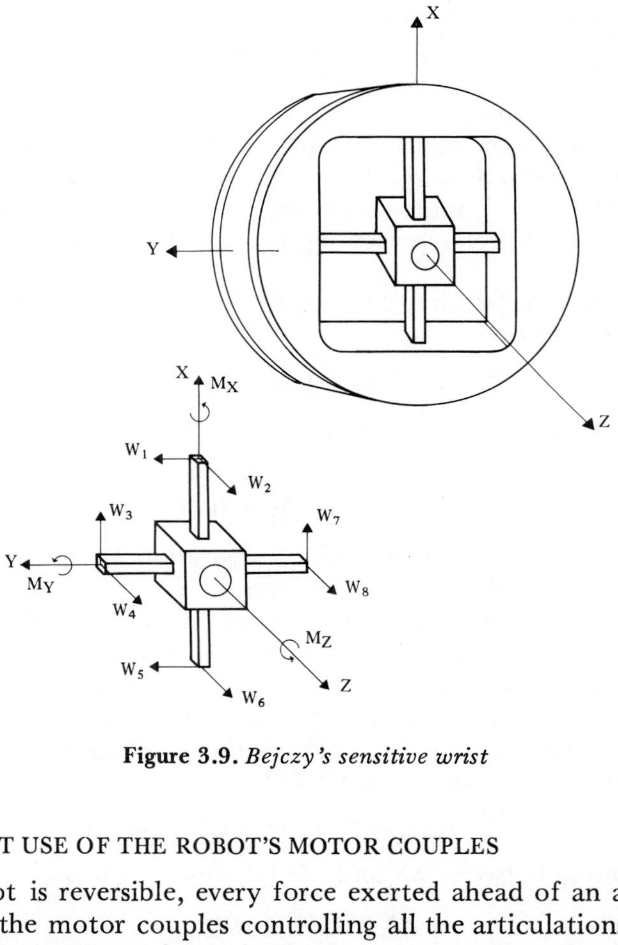

Figure 3.9. *Bejczy's sensitive wrist*

3.4.3 DIRECT USE OF THE ROBOT'S MOTOR COUPLES

If the robot is reversible, every force exerted ahead of an articulation will affect the motor couples controlling all the articulations behind it. Owing to the presence of servoing in these motors, they develop couples which will oppose movement normally provoked by this force. Measurement of difference in magnitude of these couples in the absence, and then in the presence, of disturbing force allows identification of the disturbing force. This has been used advantageously by, for example, Inoue,[14] Khalil,[15] Borrel.[16] They used this to introduce a cylinder into a bore with a very tight tolerance. Bolles and Paul[17] have used it to assemble waterpump elements driven by a Stanford arm. The great advantage of this method is that it does not need specialized sensors. On the other hand, if measurements of the forces and moments are not direct as with the sensors previously described this robot suffers from the effects of inertia, of the configuration, its load, its elasticity and dispersion of friction in each of the articulations.[11] One must therefore take detailed account in the design of the robot for each of these difficulties (Volume 1, Chapter 7), or compensate for the poor quality of measurement with a better strategy for its use.

To return to the principle of stress measurement, using variations in motor couples; this works well using the hypothesis of a perfectly reversible, rigid robot.

In the work space, R_0, the displacement increment is $\underline{\Delta X}$ and the force to be measured is $\underline{F}$. Vector $\underline{\Theta}$ is formed by the generalized variables and vector $\underline{\Gamma}$ is the vector applied by the couples to the articulations; $\underline{Q}$ is the vector representing the couples produced by the force of gravity.

It has been seen (section 2.3.1) that, in equilibrium, the equation of the robot reduces to:

$$\underline{Q} + \underline{\Gamma} = 0 \tag{3-22}$$

The reaction force of the environment $\underline{F}$ induces a couple vector $\underline{W}$ and the new balance is described by:

$$\underline{Q} + \underline{\Gamma} + \underline{W} = 0 \tag{3-23}$$

The two equations correspond to configuration $\underline{\Theta}$, the servo-systems having held $\underline{\Theta}$ in the face of perturbations.

Thus, using the principle of virtual work, one can write that the work $\underline{F}^T \cdot \underline{\Delta X}$, done by the reaction force $\underline{F}$ in moving through an infinitesimal displacement $\underline{\Delta X}$, in the work area, is equal to the work done by the active couple $\underline{\Gamma}$ in producing the infinitesimal movement $\underline{\Delta\Theta}$ of the generalized variables plus the work of the gravitational couples $\underline{Q}$ in carrying out the same displacement $\underline{\Delta\Theta}$:

$$\underline{F}^T \cdot \underline{\Delta X} = \underline{\Gamma}^T \cdot \underline{\Delta\Theta} + \underline{Q}^T \cdot \underline{\Delta\Theta} \tag{3-24}$$

The relationship between ΔX and $\Delta\Theta$ (Volume 1, Chapter 6) is:

$$\underline{\Delta X} = [J(\underline{\Theta})] \cdot \underline{\Delta\Theta} \tag{3-25}$$

where (J) is the Jacobian matrix which connects the task area to the area of the generalized variables. Then:

$$\underline{F}^T = \underline{\Gamma}^T (J)^{-I} + \underline{Q}^T (J)^{-I} \tag{3-26}$$

Note that measurment is not possible if the robot is in a singular configuration implying the reduction to zero of the matrix determinant (J) (Volume 1, Chapter 6).

Example: The arm in Figure 3.10, of type 2R—P, is in contact with its environment at point C. Using the usual notations:

$$M_0^1 = \begin{pmatrix} C1 & -S1 & 0 \\ S1 & C1 & 0 \\ 0 & 0 & 1 \end{pmatrix} \quad M_1^2 = \begin{pmatrix} 1 & 0 & 0 \\ 0 & C2 & -S2 \\ 0 & S2 & C2 \end{pmatrix} \quad M_2^3 = (11) \tag{3-27}$$

$$X_c(R_0) = S1C2(L_2 + T) \quad (3\text{-}28)$$
$$Y_c(R_0) = C1C2(L_2 + T) \quad (3\text{-}29)$$
$$Z_c(R_0) = L_1 + S2(L_2 + T) \quad (3\text{-}30)$$
$$Z_{G_1}(R_0) = \ell_1 \quad (3\text{-}31)$$
$$Z_{G_2}(R_0) = L_1 + \ell_2 S2 \quad (3\text{-}32)$$
$$Z_{G_3}(R_0) = L_1 + S2(L_2 + T/2) \quad (3\text{-}33)$$

The matrix (J) and the vector $\underset{\sim}{Q}$ take the form:

$$(J) = \begin{pmatrix} -C1C2(L_2 + T) & S1S2(L_2 + T) & -S1C2 \\ -S1C2(L_2 + T) & -C1S2(L_2 + T) & C1C2 \\ 0 & C2(L_2 + T) & S2 \end{pmatrix} \quad (3\text{-}34)$$

$$\underset{\sim}{Q} = \begin{pmatrix} 0 \\ gM_2 \ell_2 C2 + gM_3 C2(L2 + T/2) \\ gM_3 S2/2 \end{pmatrix} = (0, Q_2, Q_3)^T \quad (3\text{-}35)$$

The inverse of J is then:

$$(J)^{-1} = \frac{1}{C2(L2 + T)} \begin{pmatrix} -C1 & -S1 & 0 \\ S1S2C2 & -C1C2S2 & C^2 2 \\ -S1C^2 2(L_2 + T) & C1C^2 2(L_2 + T) & S2C2(L_2 + T) \end{pmatrix}$$

$$(3\text{-}36)$$

from which components of the reaction force may be expressed:

$$F_{x0} = [-C1\Gamma_1 + S1S2C2(\Gamma_2 + Q_2) - S1C^2 2(L_2 + T)(\Gamma_3 + Q_3)]/C2(L_2 + T) \quad (3\text{-}37)$$

$$F_{y0} = [-S1\Gamma_1 - C1C2S2(\Gamma_2 + Q_2) + C1C^2 2(L_2 + T)(\Gamma_3 + Q_3)]/C2(L_2 + T) \quad (3\text{-}38)$$

$$F_{z0} = [C2(\Gamma_2 + Q_2) + S2(L_2 + T)(\Gamma_3 + Q_3)]/(L_2 + T) \quad (3\text{-}39)$$

In equations (3-37) to (3-39):

$$\Gamma_i(F) + Q_i(F) = \Gamma_i(0) + Q_i(0) + \Delta\Gamma_i(F) + \Delta Q_i(F) \quad (3\text{-}40)$$

but: $\Gamma_i(0) + Q_i(0) = 0$ (equation (3-22)), and $Q_i(F) = Q_i(0)$ as $\underset{\sim}{Q}$ only depends on $\underset{\sim}{\Theta}$ and not on $\underline{F}$. Equation (3-40) is therefore written as:

$$\Gamma_i(F) + Q_i(F) = \Delta\Gamma_i(F) = \Delta\Gamma_i \quad (3\text{-}41)$$

$\Delta\Gamma_i$ being the difference in the couple measured in articulation θ_i for absence and presence of the force. The latter is written as:

$$F_{x0} = [-C1\Delta\Gamma_1 + S1S2C2\Delta\Gamma_2 - S1C^2 2(L_2 + T)\Delta\Gamma_3]/C2(L_2 + T) \quad (3\text{-}42)$$

$$F_{y0} = [-S1\Delta\Gamma_1 - C1S2C2\Delta\Gamma_2 + C1C^2 2(L_2 + T)\Delta\Gamma_3]/C2(L_2 + T) \quad (3\text{-}43)$$

$$F_{z0} = [C2\Delta\Gamma_2 + S2(L_2 + T)\Delta\Gamma_3]/(L_2 + T) \quad (3\text{-}44)$$

The determinant of the matrix (J) is:

$$\Delta = C2(L_2 + T) \quad (3\text{-}45)$$

which annuls itself for $\theta_2 = (K + I)\pi/2$ not being able to be negative. In these conditions, one sees that F_{x0} and F_{y0} cannot be measured and:

$$F_{z0} = \Delta\Gamma_3 \quad (3\text{-}46)$$

3.5 The use of robot-environment stress measurements

3.5.1 BETTER KNOWLEDGE OF THE ROBOT

When a robot is in contact with a supposedly rigid environment through its end effector, compliance tests can be performed by increasing the couples in the actuators and recording deformations of limbs of the robot. Then the nature of the compliance which may be non-linear, depending on configuration, can be identified and can be distorted further by parasitic actions such as those produced by dry friction. Borrel[16] thus established the elasticity coefficients of a manipulator with cable and belt transmission. In this case, the limits of limb movement were measured. When the ith articulation has reached its limit the increment ΔV_i is applied to the motor control for this articulation. By measuring variations in the motor angles and those of the couple voltage, the coefficients of elasticity introduced into the model of the manipulator can be derived. This identification is nevertheless made difficult by:

1. Noise in the measurement of couple voltage.
2. Difficulty in detecting transition from a dynamic regime to a static regime following an increase in control voltage.
3. Low value of possible increments when the articulation is in the limiting position.
4. Difficulty in detecting the limit.
5. Presence of dry friction.

This has led the author to carry out a statistical study and to obtain average values of the elasticity coefficients by trying to minimize variance.

3.5.2 BETTER KNOWLEDGE OF THE ENVIRONMENT

Following the method of section 3.5.1 stress detection can provide, in a robot whose elasticities are known, or for a very rigid robot, the identifi-

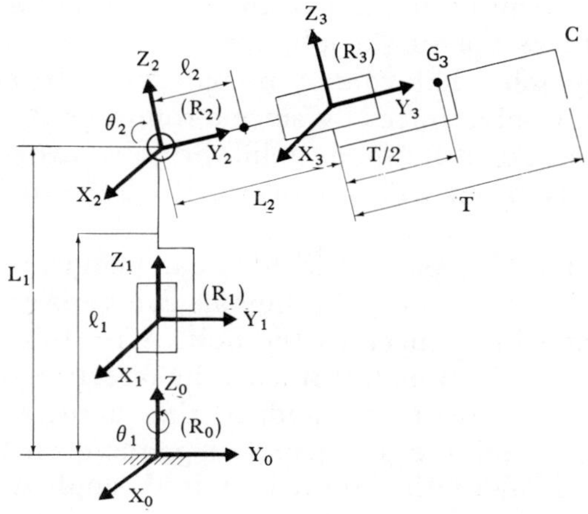

Figure 3.10. *Arm 2R – P*

cation of strength (hardness) or elasticity of material in the environment. A geometrical representation of the environment can be made equally easily. In Giraud's[18] experiment an object (a cube in his first experiment) was fitted in a compliant system equipped with gauges in such a way that when pressure was applied at a point, the position of the object of the upright component of the force exerted with respect to an identified coordinate framework could be measured.

When applying pressure in two different directions at the same point, two directions which intersect at the point under consideration were identified. A computing system equipped with *a priori* information of the object permits, starting with detailed information regarding specific points in the system, a reconstruction of the object and, in particular, the identification of its position and orientation in space. The robot can be supplied with this information before carrying out work on the object.

The aim is for the robot to discover where the object is and what its orientation is by touching it at a limited number of points on its surface. This would help solve a number of positioning problems.

3.5.3 IMPROVEMENT IN SAFETY

A knowledge of the positions of obstacles derived from stress inspection measurements can help avoid damage being inflicted upon them. This is the case in applications such as polishing, friction-cleaning tasks, glueing etc. In manipulating objects on a glass surface, stress limitation can avoid breakages.

3.5.4 AID WITH JOB EXECUTION

For safety, it is the absolute value of the interaction force that needs to be controlled. In completing a job, stress direction must be checked. There are many jobs that cannot be satisfactorily completed using positional control only, especially since a robot, unlike a machine-tool, can never be considered infinitely rigid. So jobs involving assembly or other applications often will not be carried out satisfactorily, unless stress is controlled.

This has an interesting parallel with human manual work. When a nut has to be eased onto a bolt, the human eye verifies that the nut is correctly aligned with respect to the bolt. After this, applied stress is monitored as the nut is turned. If it jams the direction of the nut can be modified slightly. Likewise, in sandpapering a surface, more or less pressure is applied on the sandpaper depending on the state of the surface and the final finish required: indeed, applied stress is continuously checked.

The classical approach in machine design is based on positional control. This influence is still very strong and in assembly problems, for example, often the first engineering attempt is to create a very precise (and therefore very heavy) machine. The use of robots, which can never be as precise as these machines because their structures employ articulations in series, puts the emphasis back onto a complementary force-position which can compete successfully with the infinitely rigid machine.

In Figure 3.11 the three possible cases of job supervision are represented; $\underline{X}$ is the position and $\underline{F}$ the reaction stress. In assembly, one seeks to neutralize stresses or some of their components; in use, one seeks to maintain a constant, predetermined stress. Positional control

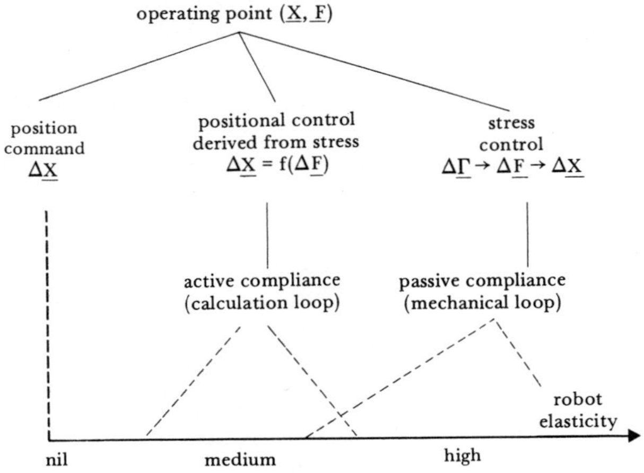

Figure 3.11. *Method of incrementation of task through interacting forces*

requires very great precision, and therefore a very rigid robot. When rigidity is less, movements must be programmed to take account of the variation in force resulting from variation in the structure of the robot. When the system becomes too elastic, it is preferable to use passive compliance. Then, in controlling stress the mechanical system distributes stress automatically. The effect is then distributed so that ultimately the robot can function effectively. This principle is employed in the RCC (Remote Center Compliance) (Watson and Drake[6]).

Another common assembly technique has been used for some time for placing eyelets, setting precious stones and other assembly processes. In these tasks approximate positional control is used and pressure is applied to obtain the final 'shift'. This technique benefits packaging rather than assembly, where a degree of deformation in the pieces can be tolerated.

Another way of approaching the general problem is to examine to what extent use can be made of a robot driven by a digital computer. Three crucial factors are:

1. Precision implicit in the design of the robot.
2. Spatial precision in the knowledge of the environment.
3. Control resolution (ie the minimum positional increment that can be achieved).

If these factors were ideal – a perfect robot model, a complete description of the environment, infinite resolution of control – one would always be able to find, whatever the elasticity of the robot, an algorithm of positional control that would allow the required task to be carried out. If one or another of the factors falls below the required standard, stress measurement is used to compensate for the resulting imprecision.

Figure 3.12 shows the mode of contact that can be employed,

control resolution	infinite									good									medium								
spatial knowledge of the environment	perfect			good			medium			perfect			good			medium			perfect			good			medium		
quality of the robot model	p	g	m	p	g	m	p	g	m	p	g	m	p	g	m	p	g	m	p	g	m	p	g	m	p	g	m
control mode for assembly or manufacture tasks	P	P	AC	P	AC	AC	AC	AC	PC	P	AC	PC	AC	AC	PC	PC	PC	PC	PC	PC	PC	PC	PC	PC	PC	PC	PC

Figure 3.12. *Mode of control as function of quality of the information in the system and resolution of control. P: perfect; g; good; m; moderate; P: possible positional control; AC: use of the possible active compliance; PC: use of the possible passive compliance. P implies naturally that AC and PC are also possible. AC implies that PC is also possible*

depending on the quality of these three factors. This naturally represents a crude assessment which serves to emphasize that use of stress is only necessary to compensate for insufficient knowledge or inadequate performance of the other parts of the system.

3.5.4.1 Example of the use of supervision stress for machine-finishing (roughing shoe uppers)[19]

The problem: A shoe consists of two main parts: the sole and the upper to which the sole should be fixed. In one technique currently used for this operation, vulcanization is used when the sole is rubber, and the injection of polymerizing materials for other synthetic sole materials. In both cases, and particularly if the upper is made of leather, sole-upper adherence is only adequately established if the side-upper surface is properly roughed ie 'brushed' in such a way that the smooth surface layers of the leather have disappeared and polymer or rubber adhesive can penetrate beyond the surface. This roughing operation, traditional among shoe manufacturers, remains uncertain if the surface coatings of the leather are insufficiently removed, giving rise to poor adhesion. On the other hand, roughing too deeply makes the leather flimsy, causing it to crack early in use. With a material as variable as leather, one can easily understand the impossibility of positional control of the roughing brush (or the milling-cutter) to achieve a good and predictable result. In practice the operation is often carried out by a worker who manoeuvres a milling-cutter and estimates by eye what he should do. He moves the milling-cutter by applying differing pressures whilst roughing, the quality of the work being monitored by eye.

Automation of such a process using a robot cannot be satisfactorily carried out (particularly if carding is carried out on already formed uppers ie not on a flat surface) without first determining the stress applied by the milling-cutter all along the trajectory that it describes.

3.5.4.2 Principle of implementation

A robot with six DOF (Figure 3.13) could control the milling-cutter in any spatial direction, to maintain a tangent coinciding with the direction needed to rough the leather. To regulate the pressure which should normally be applied to the surface (Figure 3.14), a seventh, translational DOF (T_7) is also needed.

This DOF is irrelevant to the other six but allows uncoupling of the required trajectory from the force that needs to be applied. Likewise, it might be thought that to direct the milling-cutter in space one would only need five DOF. In fact, normal F_N at the roughing surface requires, as shown in Figure 3.15, the capability of turning CAB about Z_4, to ensure the movement of the point of application of F_N to point P, so

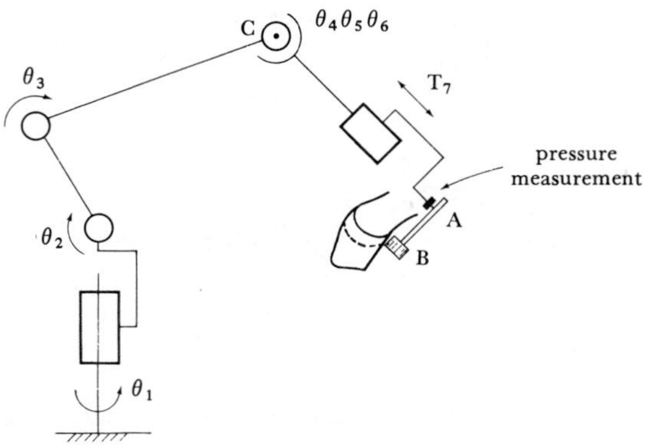

Figure 3.13. *Structure of a leather-carding robot*

that CP runs parallel with AB. The principle employed is simple. The process begins by recording, point for point, the trajectory which the milling-cutter should follow, whilst applying a pressure force of zero. After this, the force needed is applied using previous experience of the nature of the leather, the speed and the sense of rotation of the milling-cutter, its translational speed and also the part of the roughed upper. On the subsequent runs regulation of the pressure is exercised simply by controlling T_7 as a function of the difference between the necessary force and the actual force applied, measured by a bridge of suitably placed gauges. Such a system is a typical application of stress control in machine-finishing work using a robot.

5.3.4.3 Examples of use of stress control in assembly

An example in which a trajectory is generated from previous experience has already been discussed. It is around this trajectory (in use) that on-line adaptations are made, so that stress may be regulated towards a desired value at every moment. Adaptation is therefore reduced to simple programmable regulation.

This does not apply in the case of an assembly of two parts (bolt

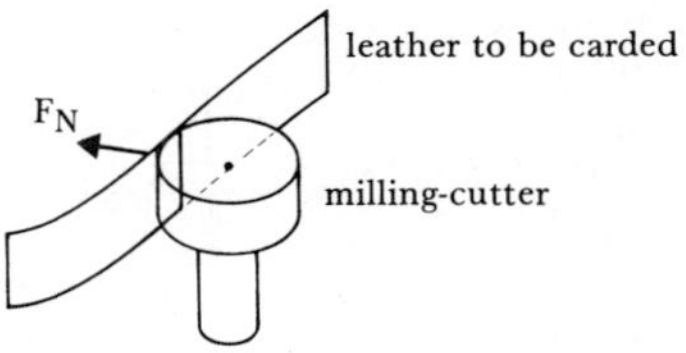

Figure 3.14. *Carding position for the milling-cutter*

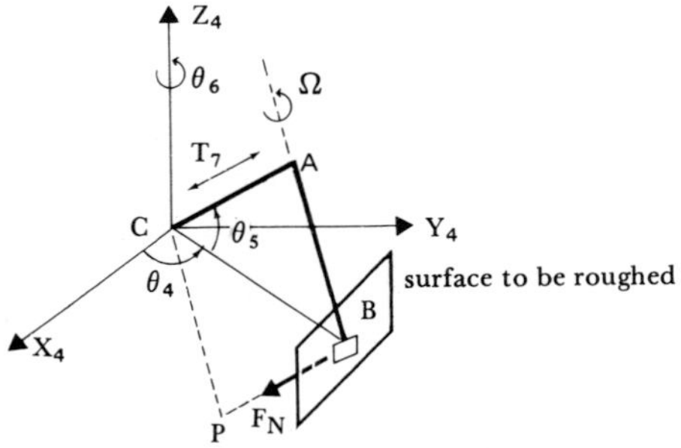

Figure 3.15. *Justification for addition of DOF θ_6*

assembly, for example). The principal trajectory might not be capable of being defined (and hence pursued) until the end of the task. In contrast with the previous example, *progress* along this principal trajectory depends on correct *detection* and the use of stress (Figure 3.16). The strategy used in such assembly work is shown in Figure 1.4; it can be varied according to the quality of the measured information.

The Nevins and Whitney method:[20] Nevins and Whitney have used, for assembly work, a strategy called *accommodation.* It consists of measuring the forces or couples in the framework of the task or the end effector and in applying fixed velocity-controlled movements in the same framework (Figure 3.17). A speed transformer (RMRC: *resolved motion rate controller*) converts the velocities in the end effector reference frame into the velocities of the articulations of the robot.

The reasoning is easier to follow if the term 'velocity' is replaced by the term 'increment' (for unit time). When the movement $\underline{\Delta X_M} = (K_F)\underline{F}$ produced on the sensor corresponds to the required increment $\underline{\Delta X_d}$, assembly is being carried out correctly. Not that the RMRC is nothing other than the inverse of the Jacobian matrix $J(\underline{\Theta})$ studied in Volume 1, Chapter 6. There are two cases in which assembly proves difficult:

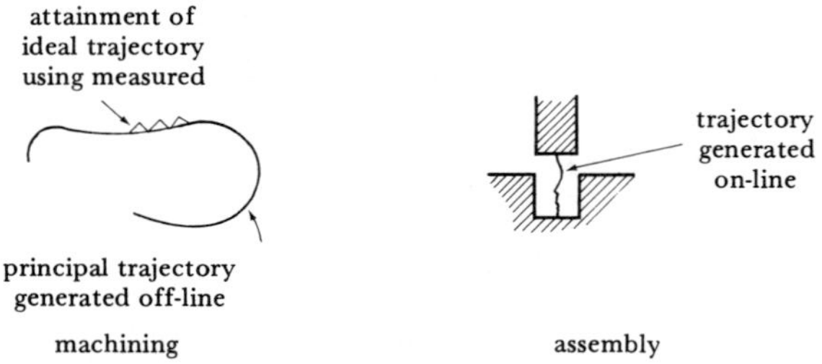

Figure 3.16. *A practical difference between machining and assembly*

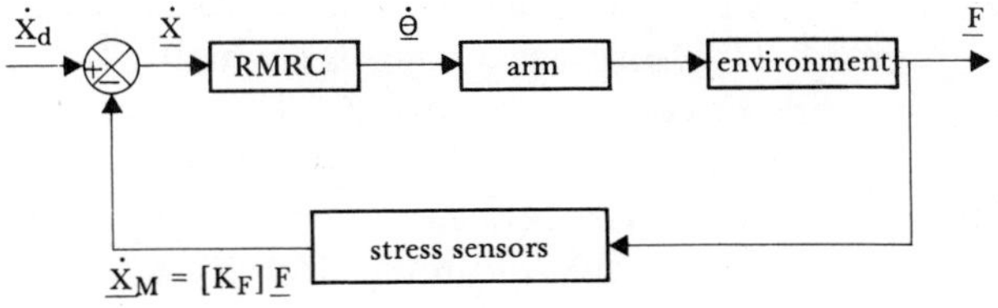

Figure 3.17. *Principle of the RMRC (Nevins and Witney*[20]*)*

1. When frictional forces are sufficient to mask the forces to be measured. Then the parts no longer come together easily and can become jammed.
2. When the two parts are in contact at two points (Figure 3.18). There again, depending upon the relationships between the dimensions of the parts and the depth to be attained, risk of jamming exists.

The solution adopted after theoretical study was to modify the compliance detector so that the cylinder, submitted to a lateral force, had translational motion without rotation. Watson[21] has carried out the theoretical study and used such a detector in an application where jamming is possible. An assembly of a cylinder of 40 mm diameter and a bore of tolerance 0.015 mm was achieved in 0.2 seconds.

Experiments of Khalil and Borrel: Khalil,[15] Borrel,[16] Inoue,[14] Goto *et al.*[8] and Bolles and Paul[17] tried to use a special stress detector added to the robot as discussed in section 3.4.3. In these conditions success of the assembly process depends upon the accuracy of the model of the manipulator allowing adequate estimation of reaction forces. If one supposes that stress has been properly assessed, then success still depends on the strategy used. If, on the other hand, measurement of stress produces excessive noise, then an increased success rate can only be maintained using a more complex strategy. Thus Khalil has achieved the insertion of a cylindrical solid of 30 mm diameter into a bore, using a H6-GZ type system with a MA23 manipulator (which is relatively flexible). Although the robot has six DOF, the author only measured the couple variation of the first three articulations (the arm). The

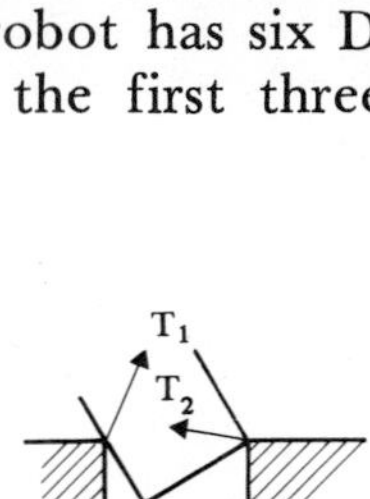

Figure 3.18. *Double contact in an insertion*

strategy was carried out in several stages:

1. Finding the bore: The positioning error could be such that the cylinder could not be inserted. The bore has a chamfer which was detected using a spiral or star-shaped movement (Figure 3.19). When the chamfer was detected, the reaction force was no longer vertical. Its orientation provided the direction of horizontal movement to be followed. Hence through passive compliance the cylinder and bore axes would then intersect.

2. Assembly: From this point, only the orientation of the cylinder, and not its horizontal position, should be modified. When the cylinder and bore axes are merged, vertical movement will ensure a fit. Modifications in orientation are related to the forces detected. Pressing vertically, if the horizontal component of the reaction force increases, the direction of this horizontal component leads to a micromovement which tends to make it decrease. This force is less than the threshold value for fitting the cylinder. Complete insertion is also verified by a stress measurement. Applying a vertical force verifies that the horizontal forces are less than the threshold and one ascertains that no further movement occurs despite the increase of the vertical force, up to another threshold. One concludes from this that assembly is complete.

The success rate is very high but does not reach 100 per cent because the effect of transmission and elasticity of the robot, between the place where the forces develop and the place where they are measured, reduces accuracy of measurement. This can have a particularly important influence on the accuracy of the detection of the chamfer.

Borrel has tried to use an improved model incorporating elasticity. This should improve the quality of stress measurement, to lift the success rate to 100 per cent. On the other hand, using a more complex assembly-algorithm the duration of the manoeuvre (finding and assembly) increases from 1-2 seconds to 5-8 seconds. This example clearly shows that the advantages and disadvantages between the rate of success and the time of calculation create a fundamental need for finding a cost-effective compromise for the use of the robot in industry.

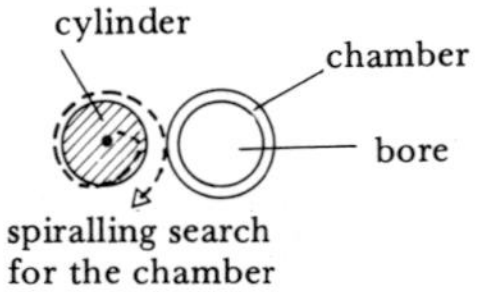

Figure 3.19. *Finding the bore*

Conclusions

In this chapter it has been shown that the mastering of stress control in environment-robot interaction is characterized by various elements:

1. Control is relatively easy if the robot is rigid enough because:
   (a) Stress sensors are simple to use.
   (b) The measurements to be carried out are small in number. Generally they concern six components characterizing a single vector force.
2. Control is very useful and even indispensable in a large number of industrial applications where positional supervision in open loop is not feasible.[22]
3. It requires the control system of the robot to be of the 'open' type (unlike currently used types) ie that it can receive information from local sources of calculation and measurement and can react to this information.[23]
4. It provides for the development of a new concept, that of auto-adaptativity, and thus enlarges the concept of regulation. In such schemes, the output is no longer a simple physical quantity, but begins to have the dimensions of *quality*, which may be improved if found lacking.

Chapter 4

Interactions involving physical contact between robot and environment: tactile detection

Tactile detection is the interaction between robot and environment such that the geometry of the environment may be recognized. This implies physical contact and requires:

1. Making the contact in such a way that both robot and environment are unharmed as a result.
2. Identifying the contact as explained in the introduction to Chapter 3.
3. Developing a strategy for scanning the environment, either in a predetermined way, or when the information acquired on-line is interpreted, the first contact having been generated in a predetermined way.

4.1 Location of tactile sensors on the robot

4.1.1 THE HUMAN EXAMPLE

Man has frequently been referred to as the 'perfect robot'. Man is so developed in the control of his movements and his understanding of the environment that he seems to possess, to a high degree, the characteristics of versatility and adaptability to his environment being sought after by robots. It is therefore natural to take man as a paradigm for these characteristics.

Man's biological tactile feelers are numerous and can be separated into several categories, as specified by Peruchon,[24] from whom we borrow, with his kind permission, the main points in the summary that follows (Figure 4.1).

4.1.1.1 Tactile safety sensors

These are essentially the *pain receptors* and *thermo-receptors*. Pain receptors have a structure identical with those of thermo-receptors, but a stimulus of intensity greater than a threshold value is interpreted by the central nervous system as a feeling of pain. They are effectively safety devices that are indispensable for the protection of the individual.

anatomical position	mechanical receptors	
	external receptors	internal receptors
epidermis	– free terminations (A and C fibres) – Merkel disks (wide band filter)	
dermis	– Meissner corpuscles (high pass) – Pacini corpuscles (high pass) – Ruffini corpuscles (low pass)	
muscle		– neuromuscular spindle – primary terminations (high pass) – secondary terminations (low pass) – Golgi organs (large band) – Pacini type receptors (high pass)
tendons		– Golgi/Mazzoni corpuscles
articulation		– Ruffini type receptor (capsule) – Golgi receptor (ligaments)

Figure 4.1. *Mechanical touch receptors in man (after Peruchon[24])*

The thermo-receptors identify 'warm' and 'cold' in man, with obvious applications.

4.1.1.2 Internal mechanical receptors

These are the equivalent to sensors that monitor position and mechanical couples which, in a robot, determine its configuration and the regulation of its actuators. One can distinguish four principal types in man.

Neuromuscular spindle:[25] This is located in the muscle itself and forms an essential element in muscular contraction. It is responsible for sensing muscular stretch and particularly necessary for controlling the cohesion of the skeletal elements and for precision of movement. The receivers possess two kinds of nerve terminations:

- *Primary* terminations that sense the speed and amplitude of the muscle's traction. In fact, their high speed of response allows them to act as acceleration sensors.
- *Secondary* terminations which behave as low-pass filters and which measure the level of continuous traction.

Golgi organ:[26] This detector is situated at the muscle-tendon junction and in the articular ligaments. It has a slow adaptation response and generates a stable output to a constant stimulus. It behaves like a force sensor and participates in the control of muscular tension by contracting muscle stretch as evident at the tendon-muscle link.

Articular receptors: These only generate signals for extreme articular positions. They do not provide information on angles as potentiometers or increment encoders do, but provide indications for extreme movements of rotational speed and tension in the ligaments.

Non-specific intramuscular receptors: These are mainly *Pacini type corpuscles* which are situated near the muscle and generate sensations of muscular tension and *Golgi-Mazzoni corpuscles* placed at the surface and interior of the tendons. (The internal mechanical receptors, as their name indicates, are not in direct communication with the environment. They allow mechanical adaptation of the configuration of the human structure to a level set by 'higher' or local neural systems. It is useful to compare the sophistication and diversity of these human systems with the simplicity of angle sensors with which robots are equipped.)

4.1.1.3 Skin receptors

These are external sensors and equivalent to those of interest in this chapter. They occur at two levels.

Outer layer of skin (epidermis):[27] There are two types of receptor in the epidermis:

1. Nerve fibres with free endings. They are classified according to diameter, conduction speed, sensitivity and the presence or absence of a myelin sheath. In the skin, fibres of the A and C type are most frequent. Type A fibres have myelin sheaths and a large diameter, low sensitivity threshold, high conduction speed (50–120 milliseconds) and extended perception field (2–5mm^2). Type C fibres are three to four times more numerous. They are characterized by the absence of myelin, small diameter, a higher sensitivity threshold and substantially lower conduction speed (2 milliseconds).
2. Terminations associated with Merkel disks, which are receptors with a broad pass band. They possess, moreover, bi-directional sensitivity which responds to deformation due to compression and shearing of the skin.

Inner layer of skin (dermis): There are three types of receptor in the dermis:

1. Meissner corpuscles[28] are numerous over the whole body, except for the fingers. They are encapsulated receptors formed by vertical and horizontal manifestations of myelinated fibres. Their ovoid axis (0.2 × 0.08 mm) is perpendicular to the surface of the skin. They are specialized in the detection of short-duration phenomena.

2. Pacini corpuscles are present in the deeper layers of the inner skin. They are formed by a central myelinated fibre surrounded by an ovoid layered and fluid-filled capsule. They are very sensitive to pressure and do not have directional sensitivity. They constitute a high-pass mechanical filter.
3. Ruffini corpuscles[25] are encountered among layers of skin tissue as well as in protective sheaths at articular joints. They are terminations encapsulated by fibres of the A type. Each corpuscle is an elongated spindle of 0.1 mm on 0.5–2 mm. In the absence of stimulation, they are permanently active, undoubtedly stimulated by tension in the surrounding tissues. They are well adapted to detect force or movement.

4.1.1.4 Conclusions

What is striking in man is first of all the *diversity* of the mechanical receptors and secondly the overlapping of their functions and their spatial influence. It is plain that the communication system existing between sensors and central nervous system as well as the handling of the information thus obtained are extremely complex, thus difficult to model and mimic.

4.1.2 MAN-MACHINE COMPARISON

One can draw useful lessons from the structure and function of human nervous receptors when considering how to equip robots with a faculty for tactile sensing the environment:

1. We are interested in *diversifying the types* of sensors to be used to equip the robot according to *function*. The functions assigned will depend on the aspect of the environment with which it will react. For example, the safety function will be provided by stress sensors (ie detection of force or couple) possessing thresholds of activation or by pseudo-contacts (proximity sensors) equivalent to thermo-receivers. The geometrical identification function of the environment will be provided using position sensors, and so on.
2. Redundancy between information derived from neighbouring sensors is a useful check of the validity of information obtained. (It monitors the quality of this information.) Use of redundancy entails an increased processing complexity (and therefore an increased calculation time) which must be kept within practicable limits.
3. Although the human body is covered with mechanical receptors, the *density* of these sensors is a function of the probability of contact between the environment and the part of the skin in question. It is probable that there is a correlation between the

sensitivity of the mechanical sensors (and therefore its type) and the density of the body zone considered. Thus, on the tip of the finger, there are up to 135 mechanical receptors per square centimetre, taking the tactile discrimination threshold up to 1.8 mm for an applied pressure as small as 0.0019 N/mm^2 (0.2 g/mm^2) (Weber[24]). The lesson to be drawn in robotics is that one would benefit from placing tactile sensors on all zones likely to have physical contact with objects in the environment, and it is natural that the end effector (which by definition will act in a privileged manner in this environment) shall not be equipped with a much denser network of mechanical receptors than the 'arm' part of the robot. (It is for this reason that 'artificial skins' have been created to cover grippers of manipulators.)

4. An essential difference between man and a fixed robot lies in the constraints which can be imposed on the environment. As an attainable space for the robot is constrained, it becomes effectively an extension of the robot. Therefore tactile sensors can be placed in this space and not on the robot itself. Take the example of artificial skin again, on which objects of interest may be placed. This skin is connected with the computer serving as the robot's brain. It becomes an integral part of the latter even if there is no physical link between the skin and the mechanical structure of the robot.

4.2 Some tactile sensors for geometrical recognition

4.2.1 ISOLATED BINARY CONTACT SENSOR

This technology can be varied. It acts as the equivalent of a switch in two positions: open and closed. Its *position* on the robot's body or in the environment is crucial to its strategic importance. Placed on the arm its sole function will be one of safety with regard to obstacles which the arm might encounter whilst in motion. These movements are complex and for each application the correct siting of the contact detectors must be determined (Figure 4.2). In most cases the environment is constructed so that the arm moves freely within it. This is why industrial robots

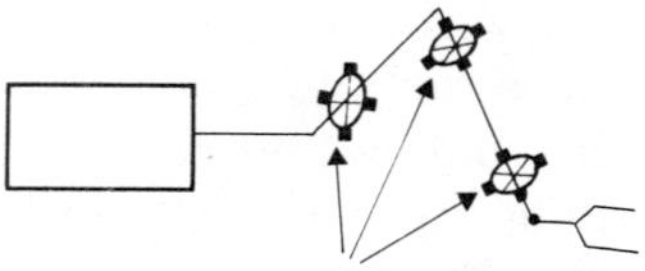

Figure 4.2. *Example of contact sensors placed around an arm*

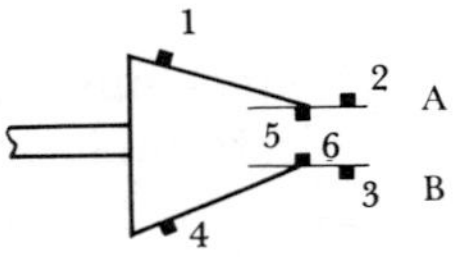

Figure 4.3. *Tactile sensors on the gripper*

generally do not carry such detectors on the arm. Placed on the end effector they become more useful (Figure 4.3). In Figure 4.3 the gripper of a manipulator is shown. If this gripper is to 'do a tour' of the whole attainable space, sensors 1 to 4 will signal safety contacts that have an influence on strategy. If sensor 1 is activated it will undoubtedly have to move downwards. Sensors 5 and 6 have very different functions: if they are activated and, if distance AB is at a minimum, it means that the gripper has not seized an object. If sensors 5 or 6 are alone activated and AB is at the maximum, it means that the gripper has moved and an object or obstacle has been encountered which could possibly be seized by the gripper. If sensors 5 and 6 are activated and $AB < AB_{max}$, it means that there is an object in the gripper.

The distance AB provides information about the objects in the gripper ie its dimensions etc. Added to pre-programmed data, this simple information can be sufficient to ensure a successful manipulation.

4.2.2 SINGLE ANALOG SENSORS

In principle, deformable systems have outputs proportional to local stress. Such sensors will therefore be able to have both static (position detection) and dynamic (force detection) applications. According to their location on the robot and their association with other sensors, their function can be varied, as can their technology. Here are three significant examples.

4.2.2.1 Peruchon's dynamic touch probe[24]

This device consists of a sensitive element capable of measuring the three components $F_x F_y F_z$ of force $\underline{F}$ exerted by the probe onto the object to be analysed. The exploratory rod (3 mm in diameter, 12 cm in length) establishes contact with the object and transmits the pressure force to the central part of a flexible, cross-shaped blade equipped with three gauge bridges which detect the component F_z of the pressure force and the two bending moments M_x and M_y of this force with respect to axes OX and OY. The system can be controlled in such a way that the pressure force is constant whatever its direction (Figure 4.4). The probe is mobilized on a system with three DOF (in Cartesian

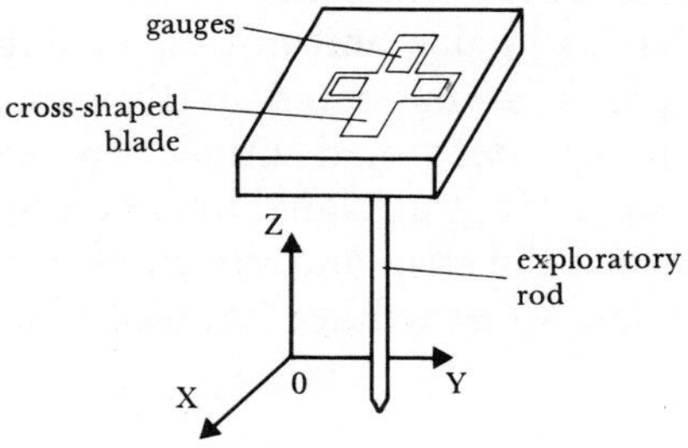

Figure 4.4. *Dynamic touch probe (after Peruchon)*

coordinates). The programming of the motion is either automatic (by program) or manual as with a joystick. Through a suitable scan, the contours of the robot's environment can be visualized or reconstructed.

4.2.2.2 Hill and Sword[11] gripper

Figure 4.5 shows one of the two fingers of such a gripper. Each finger is equipped with seven sensitive panels enabling it to detect contact of the end effector with the environment. Moreover, on the inner side of each finger there are 18 analog sensors which function as follows: each button activates a screen which obscures, as a function of the stress which it undergoes, a pencil of rays issued from a transmitting diode and picked up by a phototransmitter. This system controls the measurement and regulation of the clamping force and gives a rough indication of the form of the object in the gripper (a part of the system created by using 18 sensors which act as a matrix sensor).

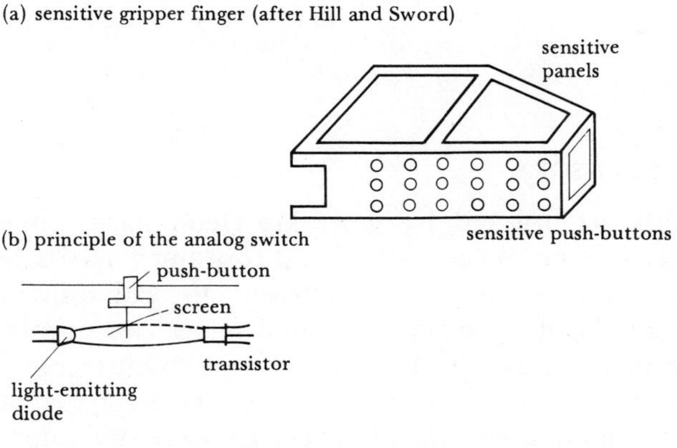

Figure 4.5. *Sensitive finger to robot manipulator*

4.2.2.3 Umetani micromanipulator[29]

This uses an analog piezoelectric sensor which Umetani used to seize objects of a few microns in size and not exceeding 100 micrograms in mass. It simply employs, as indicated in Figures 4.6 and 4.7, piezoelectric strips which are deformed under the effect of continuous tension. By sending an alternating signal between the inputs 0 and 1 the force of contact between the strip and the environment can be detected as functions of the variations of voltage induced between inputs 0 and 2.

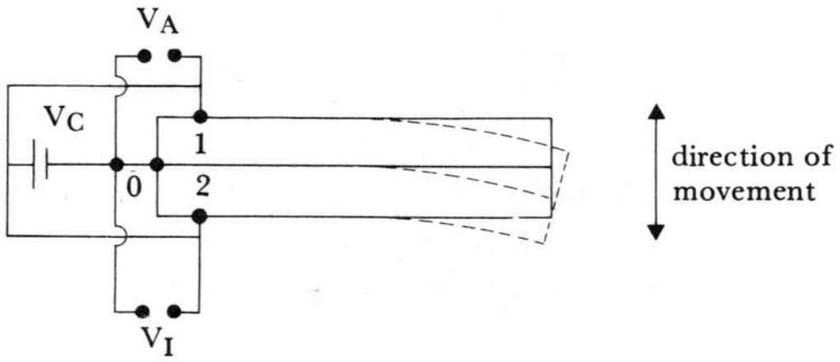

Figure 4.6. V_C*: direct voltage causing bending of the strips;* V_A*: alternating voltage superimposed on* V_C*;* V_I*: induced voltage used to measure the contact force between the strips and the environment*

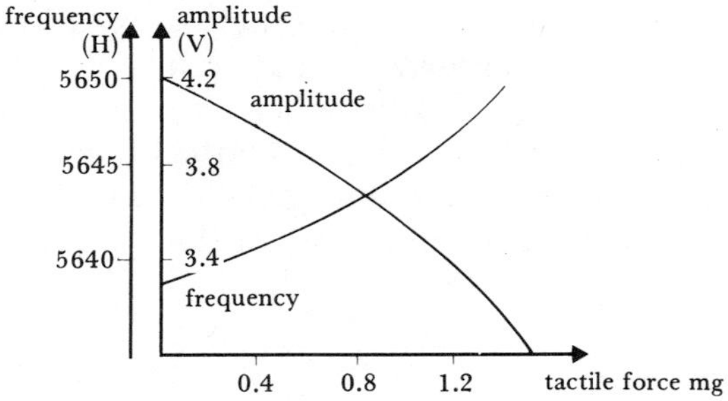

Figure 4.7. *Amplitude and frequency of* V_I *as a function of the tactile force (Umetani*[29]*)*

4.2.3 MATRIX SENSORS

These sensors are formed by grouping elementary, binary or analog devices, placed close to each other, and forming a matrix. Each elementary sensor is thus located by the line and the column at whose intersection it finds itself. If every individual feeler can supply information about force or position, the matrix network, by integration of elementary information, can produce a version of the shape of the object. This is the technique of *imprint analysis*. Here are some examples of such matrix sensors.

4.2.3.1 Sensors using piezoelectric elements[30]

The pressure-sensitive sensors are piezoelectric elements embedded in elastic polymers (Figure 4.8). The stress distribution can appear to have dimensions (X, Y). In reality, it is volumetric and the planes parallel to XY in which the imprint is identified can be accessed using the threshold voltage V_t (Figure 4.9).

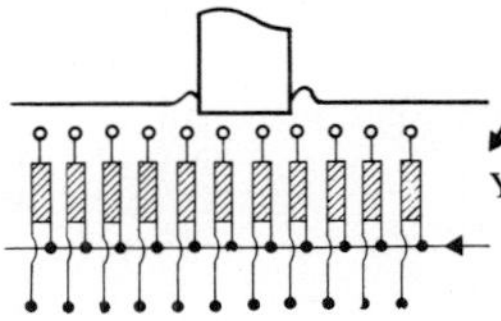

Figure 4.8. *Matrix sensor of piezoelectric elements*

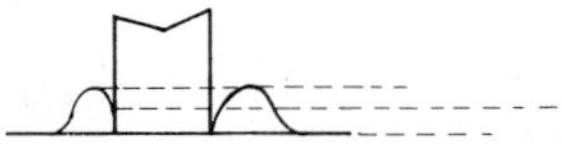

Figure 4.9. *Shape detection at several planes parallel to XY, derived from values of threshold voltage V_t*

4.2.3.2 Needle sensors

There are various models of this kind which deliver analog or logical signals. In principle they all use a matrix of rods sliding through a guide. These rods come into contact with the object to be identified. The length of the rod with respect to a reference plane is measured by various means. The differences in lengths provide the relief contours of the object. Peruchon's[24] sensor, formed by a square matrix of 64 needles, has been developed for underwater application. The positions of individual rods are measured using a differential transformer. A core of magnetic material of high permeability slides into a tube around which two windings are made, fed in opposition (Figure 4.10). Voltage is proportional to the difference in the number of turns intersected by the mobile core. It provides a movement sensor sensitive to variation in inductance.

The sensors developed by Page *et al.*[31] also deliver a signal proportional to the degree of insertion on the needles. The measurement of movement of each rod is obtained by windings similar to those used in magnetic toroids. This time not the height of the rod, but its variation in height, is measured. The system is brought into contact with the object, and (Figure 4.11) the stand moved by small increments, whilst the

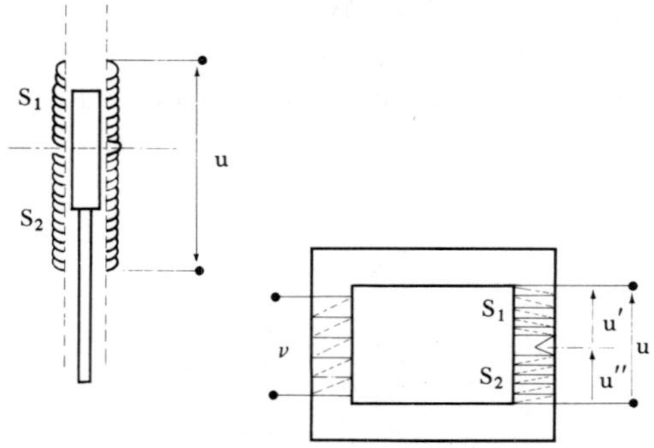

Figure 4.10. *Principle of the Peruchon sensor*

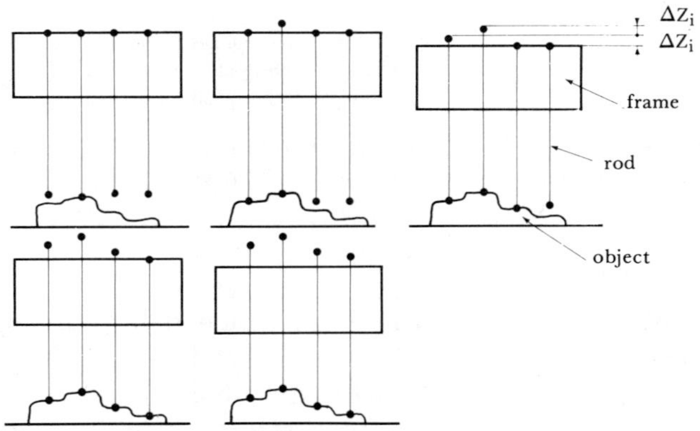

Figure 4.11. *The Page sensor: for every increment of frame movement the rods that have undergone movement ΔZ_i are counted*

motion of each rod at each increment is noted. When all rods move as the stand is moved, one cannot go any further. This allows determination of the contours of the object for different heights to be made. Goldewicht's sensor[32] delivers binary information directly. Only rods that are displaced by the object close a circuit passing through a matrix of metallic pads fixed onto an insulating stand (Figure 4.12). This matrix of 'switches' creates an image, coded as single bits. Takeda[33] has constructed on a manipulator gripper an analog device formed of two needle matrices (8 × 10) (Figure 4.13). While the gripper gap is detected separately these matrices provide a relief contour of the object gripped, by making contact in a plane parallel to Y0Z.

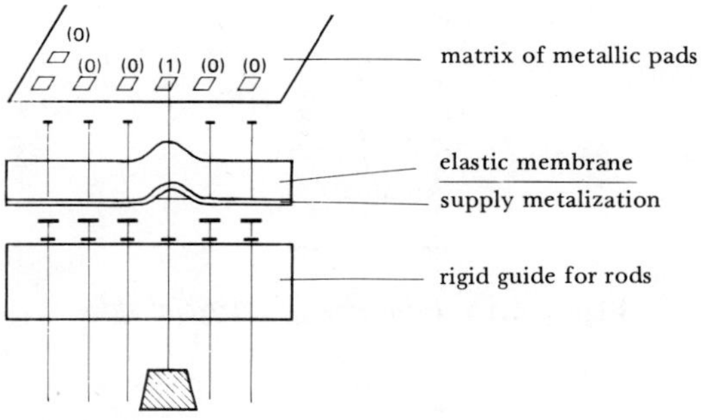

Figure 4.12. *The Goldewicht needle-transducer*

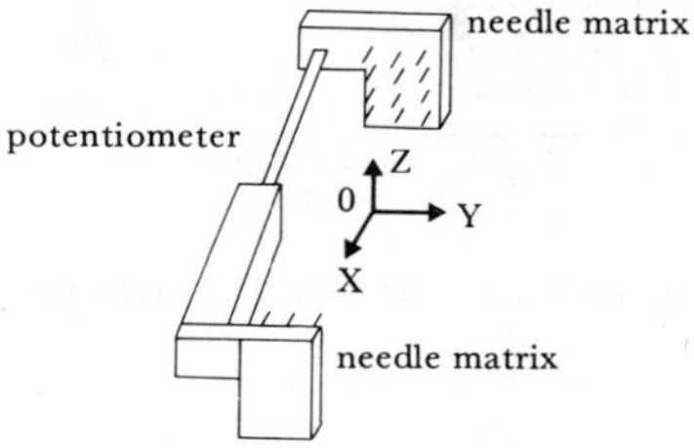

Figure 4.13. *The Takeda device*

4.2.3.3 Artificial skins

The human model of a supple, all-covering surface scattered with sensors has tempted designers who have done their utmost to create artificial, sensitive skins. These remain imperfectly developed for an ideal deformable configuration which does not suffer from hysteresis, and which could be placed in the environment as well as on the fingers of the robot. Clot *et al.* [34] in France were the first to use such sensors. Their experiment was based on the following principle.

An elastic conductive coating is placed between two electrodes (Figure 4.14). If a voltage V is applied between the two electrodes, a current I_1 flows through the coating which is of resistance R_1. If the upper electrode is pressed, this resistance diminishes and the current increases. The variation in current is a measure of the pressure exerted. The choice of coating is crucial. There are variants of this system.

In a first attempt by Bejczy,[12] metallic bands are placed above and below a plate of deformable plastic material, thus defining areas sensitive to badly defined outlines. As Figure 4.15 shows, the pressure exerted in the vicinity of a sensitive unit reduces the resistance of the deformable material and makes the current detected by this unit increase. Another version is represented in Figure 4.16.

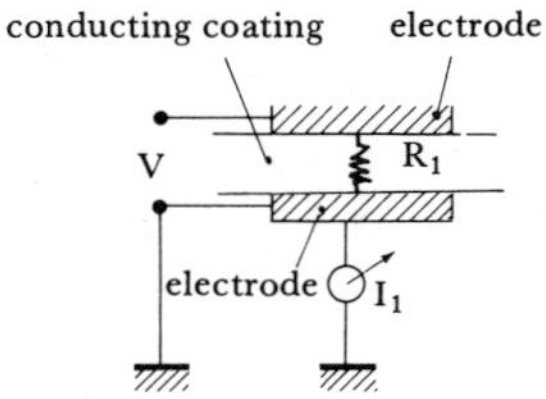

Figure 4.14. *Principle of artificial skin*

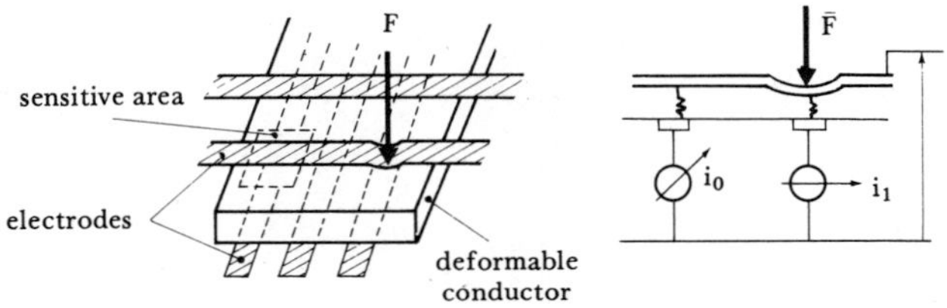

Figure 4.15. *The Bejczy artificial skin*

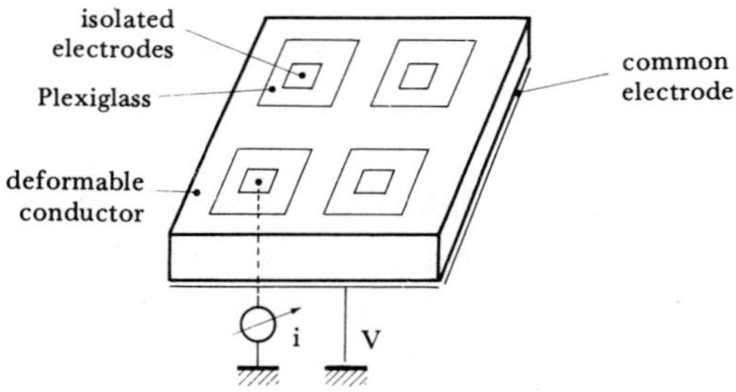

Figure 4.16. *A second version of artificial skin by Bejczy*

4.3 Use of tactile matrix sensors

Matrix sensors are used in three different ways according to their shape and technology:

1. They can form a detachable tool which the robot seizes by command and with which it explores its environment.
2. They can form an integral part of a permanently sensitive organ. The robot is thus equipped with a tactile sense. This can be a coating of artificial skin on the inside of the fingers of the gripper[35] or it can take the form of a needle matrix in the same position.[33]

3. They can be situated in the environment (a table coated with artificial skin, for example).[35]

Note that, depending on the above, there are two further ways of using the skin. In the first case, the object rests on the sensor and leaves an imprint of its weight distribution. Objects that do not rest on the sensor cannot be found, in this case. In the second case, the sensor is part of the robot: it has the advantage of being mobile and the objects are seized from 'above', providing better information. The two techniques complement each other, in fact. According to the application and the prior information, they can be used with some success. Let us note, in passing, another technique used, for example, in studying the human walk. Soles of artificial skin are placed under the feet of the walker. The dynamic imprint ie the variation in measured pressures on the soles is thus measured. When objects rest on a sensor a static picture is obtained, but if the sensor is part of the robot, dynamic behaviour may be computed and used.

4.4 Processing tactile information

The processing of tactile information is carried out in two stages. The first is concerned with measurement, primarily with such things as 'cleaning-up' data. This stage is generally called *pre-processing*. The second stage is concerned with interpretation of measurement, that is, information is provided within the framework of what needs to be known. After that, the information can be used to modify the robot's behaviour. Interaction between environment and robot clearly comes into play. When using individual sensors (non-matrix sensors) processing and interpretation are comparable with that discussed in Chapter 3 on stress detection. So they are used primarily to detect obstacles and objects of interest, to facilitate gripping in the case of manipulator robots, to improve detection for safety purposes etc. They are seldom used to detect geometrical shapes as this implies a space scan, which takes too long.

Matrix sensors, on the other hand, can be compared with low resolution optical matrices. They are used to recognize objects, which quite often necessitates the use of pattern recognition techniques: techniques whose complexity increases as the noise from the sensors increases: techniques whose complexity also increases with the difference between the quantity of the information recorded in the initial teaching stage and the quantity of information obtained from the sensor (recorded parameters exceed the measured parameters). Finally, it is noted that the complexity of these techniques increases with the number of objects to be recognized and with their similarity to one another. This is why the majority of experiments have been conducted using simple geometrical objects – spheres, cubes, cylinders and the like.

In this context the work of the following research workers should be noted: Kinoshaita *et al.*[30] using an artificial hand equipped with contacts completely or partially situated on the articulations.* Stojilkovic and Saletic[36] and Clot and Stojilkovic[37] have experimented with an artificial hand;* Okada and Tsuchiya[38] with a polyarticulated gripper; Briot *et al.*[39] with a symmetrical gripper with four fingers each comprising three finger joints* and Inoue[40] with a gripper with two fingers each comprising a matrix of 96 sensitive elements. As long as one is dealing with simple geometrical (generally convex) objects which are not representative of those found in industrial application, the mathematical method used (often with insufficient success in practical applications) can become suspect. In fact, either the method used is general in nature and becomes unstable because processing cannot be executed in real time (except when the object is made up of flat surfaces and a few elliptical sections) or the method is only valid for simple shapes such as cubes or spheres.

A more realistic example is shown in the work of Briot[35] who, using rough foundry metal parts, analysed the imprint that they left on a sensor of the artificial skin type. The sensor which was formed by $24 \times 24 = 576$ elements drew attention to some problems such as the offset voltage of the sensor amplifiers (zero had to mean a large number when no object was on the sensor), hysteresis of the conducting deformable material, a response sensitive to the position of the part because the conductor was not homogeneous. The new foundry parts had eight stable positions and the following information was required:

- The position of the centre of gravity (or its projection on the sensor).
- Orientation about a vertical axis ie location of the principal axes of inertia.
- On which side it rests.

For the first two points, the author defined a reference direction $\delta\delta'$ connected with the part, as well as a reference point in space. The recording of the imprint allows simultaneous calculations of the projection of the positions of the centre of gravity and the principal axes of inertia. Subsequently the examination of the imprint aims at discovering direction $\delta\delta'$ and the reference point within a few degrees for orientation and a few millimetres for position.

Identifying the contact face is a more delicate problem. After using a

* In these examples the sensor is not a geometrically designed matrix, but is an integration of numerous individual devices. Processing is identical with that employed for matrix feelers.

statistical approach which involved long calculation times for real time application, a syntactical approach based on regular stochastic grammars (which we will not discuss here) provided reliable identification in 0.6 second with a memory overhead of only 8k words (using a MITRA 15 computer programmed in FORTRAN). This example is very instructive. First of all it is valid whatever the shape of the object. Secondly, it shows that any object can be described *for the purpose of manipulation* by a few features (in this case: centre of gravity, contact face, orientation around vertical axis). Also the sensor, which gives even less information (an imprint) and provides more noise, nevertheless (using mathematical methods which, although sophisticated, are not time-consuming, even in 'small' computers) leads to a set of measurements in which one can have sufficient confidence to provide adequate robot control. So this represents an advance towards the solution of practical, industrial problems — an advance that will be discussed further in succeeding chapters dealing with vision.

Conclusions

A powerful interaction may be created by tactile detection if the sensors are situated on the robot. Procedures exist to ensure that contact takes place without damage. The interaction can control itself if sensors are well positioned, if the information derived is not too noisy and if its interpretation can be computed in real time. Individual tactile sensors, which can either be binary or analog, present few problems so far as signal interpretation is concerned. This is not true of matrix sensors for which techniques of pattern recognition (which will be considered later) would be needed. Matrix sensors of the artificial skin type seem, bearing in mind the factors of suppleness and integration, to be best suited for resolving the difficulties connected with tactile detection, although there are still several reliability problems to be solved.

Chapter 5

Remote interaction: proximity sensing

A factor in the viable use of robots in industry is in increasing speed. The handling of objects can lead to damage and the use of tactile detection systems can create hazards, requiring as they do physical contact. To avoid this requires use of prior knowledge of the position of the objects present in the field of the robot and a need for appropriate trajectory planning. One way of overcoming this has led to a practical consideration of *remote sensing* using 'proximity' detection. Using this, danger can be sensed early enough to avoid damage by stopping the motion of the robot or by changing its direction of movement.

Unlike the tactile sensors discussed in preceding chapters, there is no interference between contact detection and the reaction *forces* of the environment. On the other hand, the position of a contact is often known with a lower precision than that supplied by a sensor detecting physical contact.

5.1 Categories of proximity detectors

To obtain information at a distance an object needs to emit a signal (or create a field). The active proximity sensor (and here the camera, which will be considered in later chapters, is not included) is best used in the following circumstances:

(a) The object naturally transmits a signal (eg radioactive material).
(b) The object is equipped with its own transmitter.
(c) A signal is transmitted to the object and then received after reflection. This signal can be of natural origin (eg reflection of ambient light) or can be artificial.

If the first two methods are used, the feeler is a *passive* receiver. This is also true for detection of a signal of natural origin as in (c). If the signal is of artificial origin it indicates that there is an artificial *transmitter* as well as a *receiver*. When these two devices are placed on the same sensor, an *active sensor* is created.

How can an object emit identifiable, natural radiation? This is easily answered from physics in cases where radiation might be radioactive

(α, β, γ rays), luminescence (in the visible spectrum), infrared or electromagnetic radiation of all kinds. One could also include the creation of magnetic or gravitational fields that could be detected. However, outside very specific applications some of the radiations are so weak that they may not be detected.

Objects of interest in the robot's environment can carry a natural transmitter only when the object is deliberately marked, as for example when radioactive or luminescent paint is used. The uses, for example, of an artificial device such as an electronic transmitter attached to the object would be limited in practice.

But natural transmissions such as luminescence, gravitational or magnetic fields can be utilized. These fields are altered by the presence of objects in them and this property obviously can be used to provide methods for detection. Cameras using visible reflected radiation in the infrared and near ultraviolet parts of the spectrum can detect objects. Also sensors, formed by bridges whose branches contain elements sensitive to magnetic fields, show an imbalance in the vicinity of a metallic mass of strong magnetic permeability and provide yet another method of detection. The active proximity sensor which, as we shall see, turns out to be optimal in many cases, can be used:

(a) To control the transmitted signal, as a function of time.
(b) To transmit a signal of natural origin not normally present in the environment. This favours its detection.

5.2 Some types of active sensor[41]

One type of sensor uses the detection of deviation of an *air jet*. This has not been used often in the field, but has been tested in the laboratory. Although simple in principle, and easy to use in practice, its principal faults lie in its short range and need for compressed air.

Sensors based on *ultrasonics*[42] have to be mounted on special cameras because they are both cumbersome and fragile. Their range of application makes them useful for distance measurement – they depend on measurement of time between a transmitted and reflected pulse. Use of the Doppler effect provides a basis for *speed measurement*.

Radio-frequency waves are of limited potential at the moment. They have to be ruled out on the grounds of complexity and cost.

The last category of active sensors are those which emit a luminous radiation. Their range, from a few centimetres to some tens of centimetres, makes this method of interest to robotics. Moreover, the luminous source can be located far from the actual sensor if glass-fibre optics is used. *Infrared radiation* could be used in a large number of applications. In Figure 5.1 the principle of the optical sensor is demonstrated.

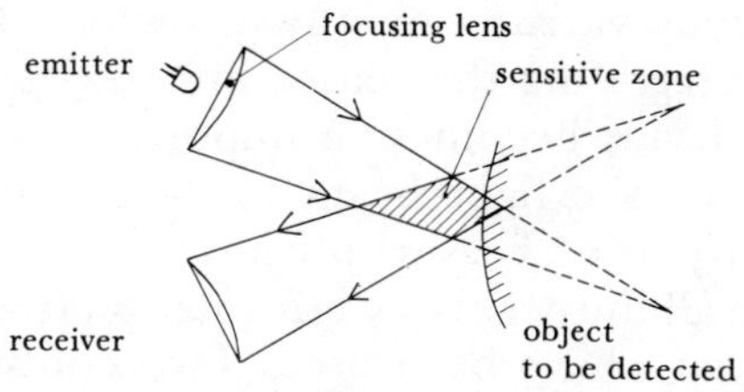

Figure 5.1. *Principle of a photoelectric proximity sensor*

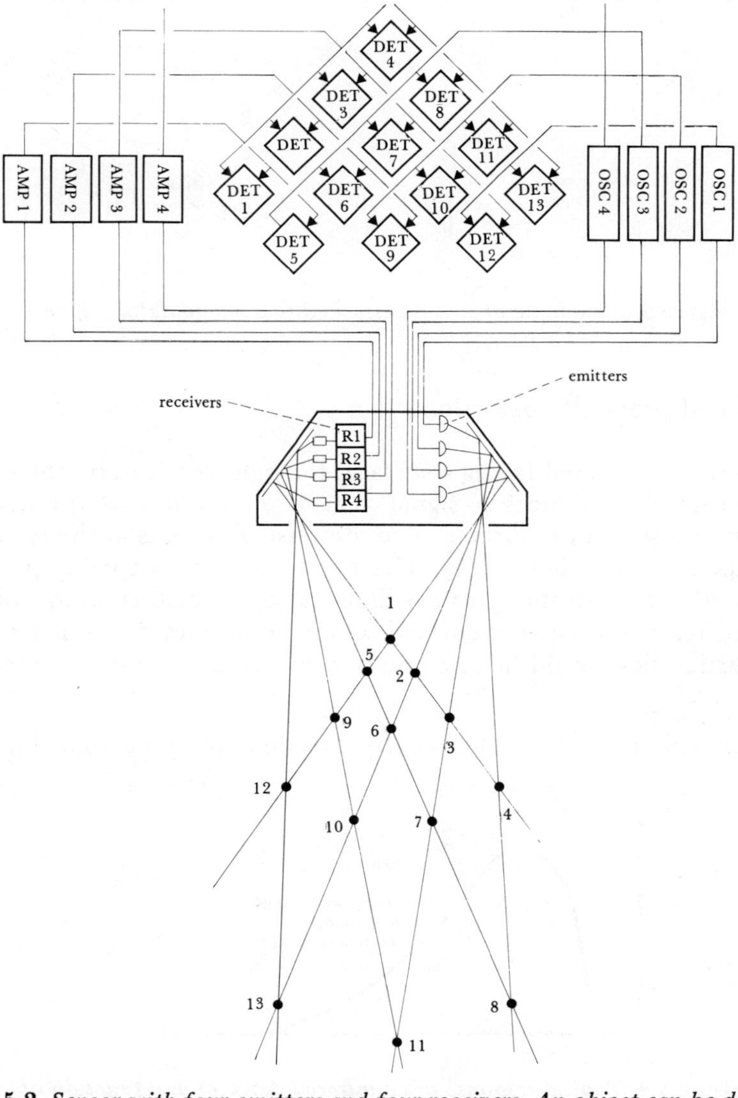

Figure 5.2. *Sensor with four emitters and four receivers. An object can be detected over 13 zones, surveyed in the 'all or nothing' mode*

There are numerous variants on this principle. In robots it is most often a matter of integrating the sensor into the gripping tool,[12] when the volume of the device becomes an important factor. There are also problems in design of the light beam to be considered eg the use of lenses can be aided by using Fresnel prisms.[43]

Finally, to obtain all the necessary information from the environment, *multiple sensor systems* have been used. Two contrasting examples are the multiple sensor such as that shown in Figure 5.2, which was developed at the Jet Propulsion Laboratory,[44] or the integration of individual sensors into various strategic locations in the robot (Figure 5.3).

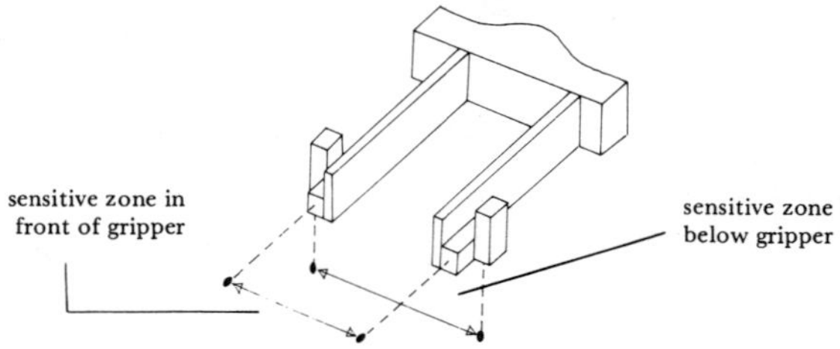

Figure 5.3. *Example of sensors attached to a manipulator gripper*

5.3 Infrared proximity detection

If a sensor is positioned facing a surface, the light received by the detecting photodiode produces a signal that is a function of the distance between sensor and the surface. The response pattern is in the form of a bell-shaped curve (Figure 5.4). The pencil of rays not being precisely defined means that this curve will be slightly distorted for objects which differ from a plane surface. Given a smooth reflecting surface three difficulties would be encountered in making a distance measurement:

1. The whiteness of the surface (by trial and error) has to be known.

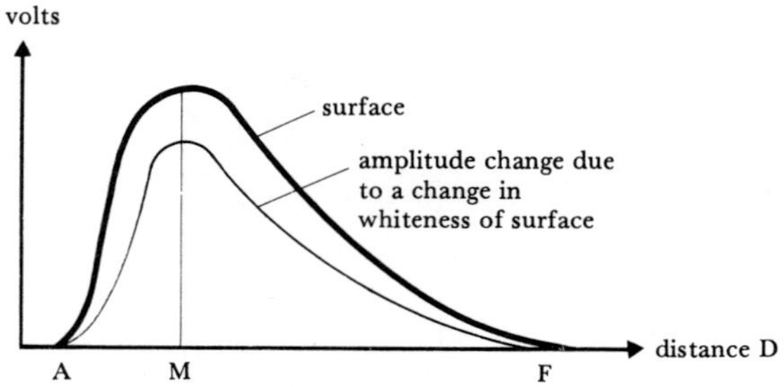

Figure 5.4. *Typical response of an infrared detector as a function of distance from a surface*

2. Except for the singularity at the top of the curve, the same signal can denote two possible distances.
3. The axis of the sensor must be normal to the surface.

In applications using manipulator robots, sensors are often placed on the gripper in pairs. Because the gripper is mobile, nothing ensures that the axes of the feelers are normal to the surfaces of the objects which are in unknown positions. The curve obtained in practice will no longer be as in Figure 5.4. This is why these proximity sensors are mostly used for the detection of the presence of an object in the volume, scanned only in rarer cases to measure distance, and, exceptionally, in recognition work. Note that, by using several stages the simple 'presence-absence' signal can allow one to 'surround' the object ie to position the system of sensors in an optimized direction with regard to the object. This is used to advantage in manipulator robots which need to grip objects automatically.

5.4 Object detection

The general example in Figure 5.5 shows a gripper with two fingers. The end effector of a robot is equipped with two proximity sensors. This gripper is mobile and the presence of an object, a cylinder, for example, is detected by scanning the space in a plane parallel to that where the object rests.

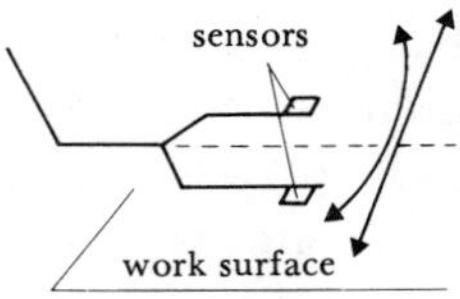

Figure 5.5. *The two-sensor gripper scans by rotation and translation in the plane parallel to that of the work surface*

The signal received in translational movement by one of the sensors provides curves such as those shown in Figure 5.6. The sampling frequency used does not exceed a few dozen hertz.

Adequate detection depends upon obtaining a meaningful curve of detector signal s(t) against distance x(t). The problems that arise are:

(a) Noise produced by the sensor and other components.
(b) The signal s(t) will include the effect of disturbances produced by the presence of other objects, and will depend on the shape of the object,
(c) s(t) is a signal whose frequency characteristics will depend on the speed of movement of the feeler in front of the object.

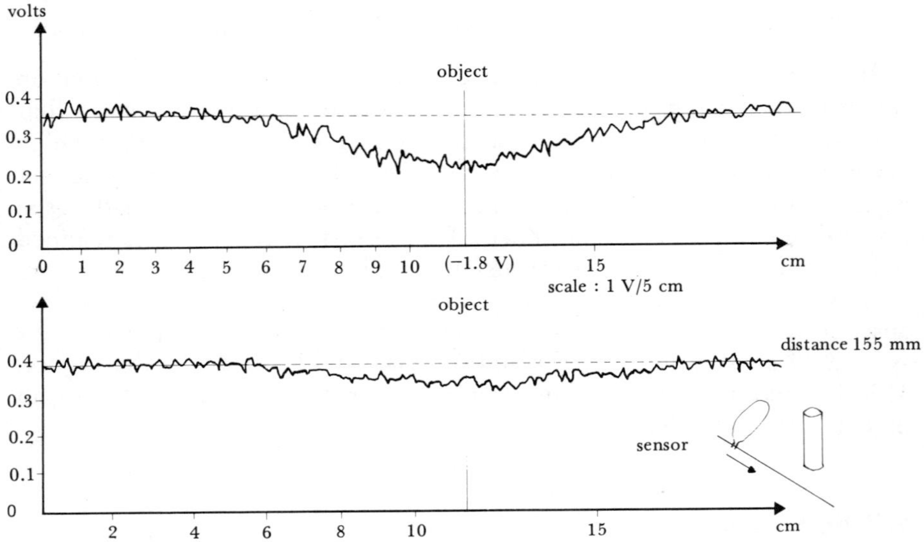

Figure 5.6. *Example of the response of a sensor moving in front of a 18-mm diameter cylinder placed at 30 mm and 150 mm from the line followed by the sensor*[45]

To detect the object from the characteristic curve, there are numerous possible techniques which are reviewed by Ruget and Espiau:[45]

1. Low-pass filtering in the form of an integrator. This provides a 'smoothed' value of signal which varies as a function of x(t). A variation in this average value indicates the presence of an object.

2. The technique of signal sampling at regular intervals. The signal, as before, is integrated and I compared with an arbitrary limit λ. If $I > \lambda$ an object is detected. The integral is set to zero at each interval T.

3. More sophisticated filters that rely on spectral methods, probabilistic techniques or methods that extract signals from noise, for example.[46]

4. Detection by change of model. This consists of modelling the signal delivered by the feeler in the absence of an object. The model will vary in the presence of an object. Such a model for these signals needs some consideration. A possible approach consists of looking for an iterative, linear model of the form:

$$s(n) + a_1 s(n-1) + \ldots + a_p s(n-p) = b_0 e(n) \qquad \textbf{(5-1)}$$

in which e(n) is a sequence of independent, Gaussian, random variables taken from the set (0, 1). From equation (5-1) is derived:

$$s(n) = -\sum_{i=1}^{p} a_i s(n-i) + b_0 e(n) \tag{5-2}$$

The coefficients a_i and the order p of the system must be determined. For coefficients a_i:

$$\underline{a}_n^T = [a_1 \ldots a_p] \tag{5-3}$$

$$\underline{s^T(n)} = [s(n-1), s(n-2), \ldots, s(n-p)] \tag{5-4}$$

Equation (5-2) takes the form:

$$s(n) = \underline{s^T(n)} \cdot a_n + b_0 e(n) \tag{5-5}$$

$$\underline{a}_{n+1} = \underline{a}_n \tag{5-6}$$

The identification of $\underline{a}_n$ is a classical automation problem which can be resolved with the help of the related Kalman filter. P must not be too high and yet it must be representative of the system. Akaike's criterion connected with the average quadratic error can be applied. It is expressed by finding the minimum value of the expression:

$$A_m = \log(\epsilon_m) + \frac{2m}{N} \tag{5-7}$$

in which N is the size of the analysis window and ϵ_m is the average quadratic error of:

$$\epsilon = s(n) - \sum_{i=1}^{p} a_i s(n-i) \tag{5-8}$$

$$\epsilon_m = \frac{1}{N} \sum_{j=1}^{N} \epsilon^2 \tag{5-9}$$

When m is increased, A_m passes through a minimum which is adopted as order p of the model. Given the model, it is necessary to define the discriminant function that allows detection of change of model. One can use several types of distance, for example:

$$\delta = \sum_{i=1}^{p} |a_i - \underline{a}_i| \tag{5-10}$$

a_i being the parameters of the model and $\underline{a}_i$ the parameters estimated from the measurements.

5. Detection based on statistics of sequences (Moskov chains) can provide a useful method, involving simple algorithms compatible with real time control problems in small computers whose complexity is limited by prior knowledge of the system.

The principle is the following: in the sequence of measurements one seeks to detect an increase followed by a decrease. At n_i^0 suppose that p measurements have been made:

(a) q measurements are read following measurement i.
(b) The existing p + q values are classified in decreasing order.
(c) Among those the highest p + 1 values are kept.
(d) The number of conserved values that come from the q new measurements are counted. This number is written as K.
(e) K(i) is tested by comparing for example K(i) − K(i−1) or K_{max} − K(i−1) against a threshold. This method requires the control of only two parameters ie p and q. It is, moreover, highly insensitive to noise. In adjusting p, a window is moved, avoiding the problems of memory limitation.

6. A related method consists of studying the *variation of the number* of measurements in excess of a limit, using a window shifted at every step. Consider a window comprising P points and suppose that A_i is the test function:

$A_i(P, \lambda)$ = number of measurements y_j so that $y_j > \lambda$ is the interval $j = i, i + M - 1$.

Calling

$$A = \max_{K \leqslant i} A_i \qquad (5\text{-}11)$$

One can decide that the object is present if:

$$A - A_i \geqslant \gamma \qquad (5\text{-}12)$$

This method is very simple and effective if one knows how to determine P, λ and γ, which can be done by simulation, by taking as a criterion the false alarm rate, non-detection rate and average delay in detection.

In the interaction between a robot and the environment the problem of correct execution is essentially a problem of strategy based on the information derived from the sensors. As a result, extraction of this information is a local problem (which is best dealt with using a specialized processor) which requires very short response time and provides reliable results. Without this one must rely on higher computational powers in the robot using probabilistic correcting algorithms which are excessively complex. This is why the last two methods of detection described are simple and reliable (given adequate trial procedures) and are to be preferred to the more sophisticated, and ostensibly superior, methods.

5.5 Measurement of distance

It is essential that a sensor should be able to indicate distance. How does one quantify such a measure? A manipulator robot must be able to avoid obstacles: it is enough to detect them. However, it must also be able to seize objects and it is then necessary to estimate sensor-to-object distance. This information is provided, in general, by means other than proximity sensors. Distance may be estimated from the response curve in Figure 5.4. This curve:

(a) Is non-linear and non-monotic.
(b) Varies with the object from which it derived its form, its colour and its orientation with regard to the sensors.
(c) Has varying noise levels.
(d) Varies from one sensor to the other.

In these conditions, precise measurement of distance can only be carried out by bringing together a number of parameters in both the learning and measurement stages. Some of these constraints are compatible with general robotic practice. It is the same for the use of the same sensor and the same objects. Obviously these last two constraints are not sufficient and, moreover, in their work Ruget and Espiau[45] had to restrict themselves to the relative directions between the surface of the object and the axis of the sensors, leaving vital parameters constant between the learning phase and the measurement phase. This assumes that the object is known, that it has a smooth surface and that one knows its spatial orientation. In particular applications (getting a tool perpendicular to the plane of a pre-orientated piece, for example), the method of distance measurement suggested could be useful. The estimation problem is a classical filtering problem, which can be resolved by linearization of the first- or the second-order recurrent equations:

$$\left.\begin{aligned} X(K+1) &= X(K) + T(K) + V_1(K+1) \\ \lambda(K+1) &= \lambda(K) + V_2(K+1) \\ Y(K+1) &= \lambda(K+1)\, f[X(K+1)] + W(K+1) \end{aligned}\right\} \qquad (5\text{-}13)$$

In these equations $X(K)$ is the object-sensor distance at instant K; $\lambda(K)$ is an instantaneous affinity factor; $T(K)$ is the movement increment of the sensor at instant K; $f(X)$ is an 'average' response of the sensor with regard to a plane; $W(X)$ is the noise in the measurement and $V_1(K)$ and $V_2(K)$ are the uncertainties in the increments of movements. $X(K)$ and $\lambda(K)$ have to be estimated. The authors show experimentally that, in their case, a Gaussian filter of the second order gives the best results.

5.6 Recognition of objects

It would seem from what has gone before that one will require even more constraints on the relative positions of the feeler and the object than in the estimation of distance to obtain object recognition, and that applications will be extremely specific.

5.7 Environment-robot interaction in automatic gripping as monitored by proximity sensor

A classical application of proximity sensors used without an articulated robot is represented in Figure 5.5. When the gripper has been brought, by application of a program or any other method, into a zone close to the object to be seized, its movements are going to be the result of the comparative measurements made by the two sensors; the aim of the operation being to bring the gripper's axis (which we will suppose to be horizontal) to pass through a median plane of the object. After this the gripper needs only to be moved along the axis by using, for distance control, either a third proximity feeler situated at the base of the gripper, a contact sensor, or another means.

The centring signal used is the difference $\epsilon \mp V_1 - V_2$ in the signals picked up by the two sensors. This difference has the form indicated in Figure 5.7. It changes as a function of feeler-object distance d.

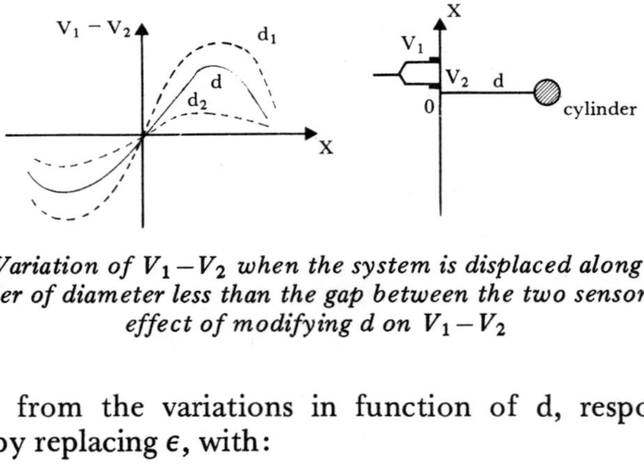

Figure 5.7. *Variation of $V_1 - V_2$ when the system is displaced along 0X in front of a cylinder of diameter less than the gap between the two sensors, and the effect of modifying d on $V_1 - V_2$*

To break from the variations in function of d, response can be normalized by replacing ϵ, with:

$$\epsilon' = \frac{V_1 - V_2}{V_1 + V_2}$$

In these conditions the centring will be perfect when $\epsilon' = 0$ (with regard to a cylinder). It is therefore simple to generate a procedure for the movement of the gripper servoed on the value of ϵ' with a break in the value where ϵ' changes sign. If the object is not cylindrical the change of sign will be produced for a position which will not correspond to a perfect centring (Figure 5.8).

This does not present problems in so far as the opening of the gripper is such that, by following the detected centring line in the direction of the object, it 'comes back' into the gripper. With this procedure a gripper can be servoed to the pursuit of a mobile object, as long as the speed of

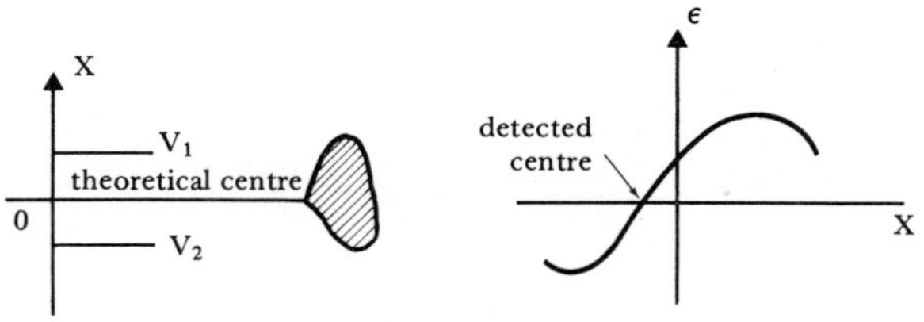

Figure 5.8. *Change of sign for a centring operation different from the theoretical one*

movement of the object is compatible with the response time of the feeler and robot system.

Conclusions

Unlike the cameras that will be studied in the following chapters, proximity detectors are not intended to recognize objects (although this can be done theoretically if proper measurements of sensor-object distance can be made for a large number of points on the object) for we have seen that use of deformation of the sensor's response curve, obtained by scanning for different orientations with regard to the objects, only allows its recognition in exceptional cases. Proximity detectors essentially allow (as their name indicates) one to 'surround' an object ie to determine a volume of the space in which the object is found. The use of several associated sensors as a reliable method of detection, and an adequate mobility strategy of the sensors, effectively reduces this volume until it is effectively a small 'box' around the object (Figure 5.9).

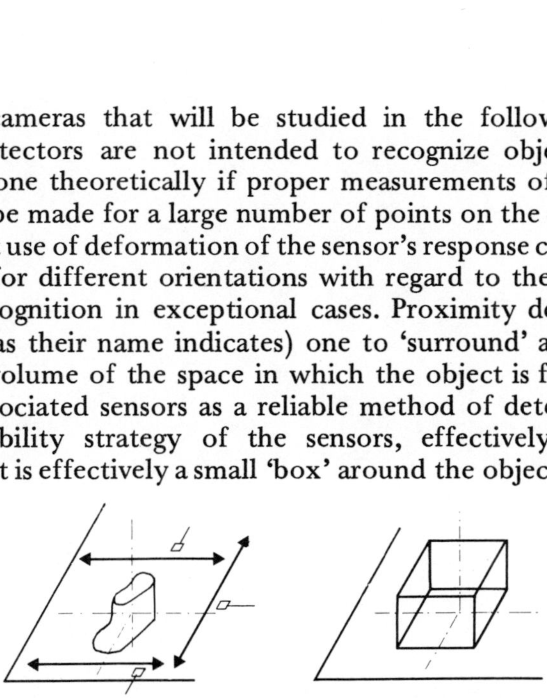

Figure 5.9. *With three movements of a detector it can be seen that the object fits into the box on the right*

This positional information is extremely important for the robots which interest us here. It allows avoidance of collisions and also detection of the object using another method to seize it safely. The most advantageous technique therefore seems to consist of recognizing the nature and the approximate position of the object using a camera, for example; when this information has been obtained the robot moves towards the object and its gripping behaviour is controlled by the information produced by the proximity sensors.

Chapter 6

Remote interaction: artificial vision*

6.1 The nature of sight

Sight and perception are very complex subjects. They are concerned with the transmission and interpretation of information received by our visual sensors – the eyes. There are three main schools of thought concerning the nature of sight:

1. The school represented by Gibson,[47] which claims that the interpretation of what our eyes receive is 'direct' or 'intuitive'. Gibson maintains that perception is concerned with the passive reception of information, derived from the ambient distribution of light, and that what we see is determined by stimulus – shapes such as texture gradients or parallaxes created by movement.[48]
2. The school which regards eyes as transducers supplying signals that 'improve' perception. Whilst perception, according to Gibson, is an impression of our surroundings, providing direct information, for the second school perception forms patterns of signals triggered off by an image of the world derived through the eyes. Then, since in this theory perception is no longer a direct process it can give rise to errors caused by a malfunction of the transducers or the transmission between sensors and brain.
3. Finally there is the school that believes, with Helmholtz,[49] that perception is related to the world in an even more indirect way. It is related by an interaction between the signals, originating at the sensors and transmitted by the nerves, and data already stored in the memory. Perception is then an extrapolation of data already present. Perception is therefore a much richer picture than the information received by the sensors alone. It too is prone to errors and confusion caused by misinterpretation.

* Since the original publication of the French text, vision in robotics has become an area of intense research activity. Consequently some of the statements found in this chapter may seem pessimistic, in particular with respect to the technological modelling of human vision in a cost-effective way. Aleksander *et al.*[50] may provide a useful reference. In this one notes that machines now exist that can recognize patterns at high resolution, at the rate of about five images per second and at costs that are a small fraction of the mechanical part of the robot.

Helmholtz's hypothesis would appear to be the most appropriate for robot systems, although it is not known exactly how the brain works in selecting patterns and making inferences. It is nevertheless instructive to look at how the human brain would function under this hypothesis.[51] Consider an object, a book, for example:

(a) The object *emits signals* ie reflected light.
(b) These signals are picked up by the light-sensitive receptors in the eyes, the cones and rods. The light signal is converted to a neural signal in the optical nerve.
(c) The signals transmitted to the brain are processed in an *associative* manner to obtain a *description of the object*. The rules of association are made by means of criteria *already present*. Thus an *image criterion* is obtained ie a representation of the object. The formation phase of the image criterion is called the *learning phase*.
(d) The memory of the brain already contains image criteria which have been acquired by learning at various stages in life. These image criteria are called *image models*. The brain compares the image criterion of the book with the image models. This is called the *recognition phase*.
(e) When there is coincidence between an image criterion and an image model, the image criterion triggers off a nervous response (Figure 6.1).

The power of the human brain lies in its ability to recognize an object or analyse a scene very quickly and independently of conditions and of observation. Even if the model described above is feasible, nothing is known about the rules governing association, the criteria and the inferences that are related to them, or anything about the nature of object description. There appears to be no artificial mechanism that allows for recognition, independent of the conditions of observation. Visual perception, then, is a poorly understood and complex pheno-

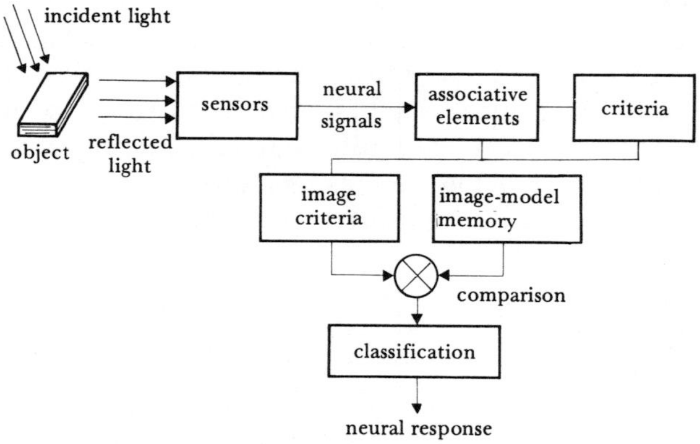

Figure 6.1. *Hypothetical functioning of the human brain in visual recognition*

menon (in man and in animals), and if it is required to equip robots with this quality great difficulties may be encountered.

Whatever the method used an artificial vision system designed for a robot must always contain:

- One or more signal *sensors* with variations dictated by the scene or by the objects of interest. In addition, the signals used need not necessarily be optical in nature (eg ultrasonic images might be used).
- The images have to be *converted* in such a way that the signals derived from them are able to actuate the equipment which handles them (eg conversion of the optical signal into an electrical signal is the most common).
- The image then has to be transformed so as to extract only the information needed. This stage, identical with the description of the image in the previous model, is the *representation* stage of the image.
- Then, to this representation various processes have to be applied which are going to provide the descriptive information related to the images. This is the stage of *extraction of the relevant features* or *pertinent information*.
- The characteristics obtained can be compared with those that have been recorded in the learning stage. This then represents the *recognition* phase.
- Recognition can be complete, partial or non-existent. This can be interpreted in two ways: either as a probability of an entire recognition, or as a partial recognition that certain elements of the scene have been left out. Whichever is the case, a decision to act, following recognition, has to be made.

It is immediately noted that the structure of an artificial vision system

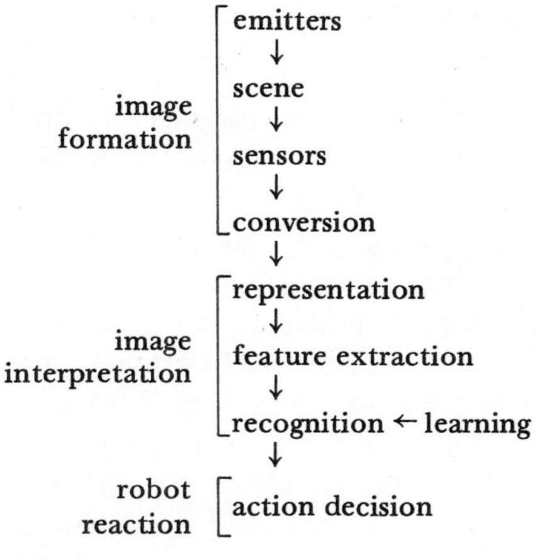

Figure 6.2. *Phases of a robotic artificial vision system*

must be related to the information one seeks to obtain through vision. As it is impossible to create a device comparable with human vision, the solutions will vary with the degree of versatility required. The phases shown in Figure 6.2 will vary with the technological means or the equipment used. To deal with the matter fully would involve use of artificial intelligence — the subject of a later volume in this series.

6.2 What is an image?

The image is presented as an input to a sensor. In this chapter the images considered will be *bi-dimensional.* They can be represented as a surface in which each point (X, Y) detects an intensity g(X, Y) (not necessarily *luminous* intensity since the radiation need not always be optical).

The function g(X, Y) represents an image. Mathematically this will be, for a real image, a continuous function or a function possessing finite discontinuities, positive if one supposes that g(X, Y) = 0 corresponds to zero intensity.

A *tri-dimensional* image can be represented by [g(X, Y), r(X, Y)], r being a function which associates the distance between the point (X, Y) on the sensor and the transmitting point of the object. This distance function is not necessarily provided by a direct distance measurement.

6.3 Pattern recognition as applied to robots

Pattern recognition is not a clearly defined area of study. It is associated with identification problems encountered in automation.

The methods developed in pattern recognition have been developed within a different context from that encountered in robotic vision, most often being used to analyse photographs. So such methods are not always immediately applicable to robotics. The main difficulties are:

(a) In robotics image analysis in *real time* is imperative. It is this analysis which will provide the information necessary for the execution of the robot task. Real time, in industrial use, requires image analysis completed in a fraction of a second, ideally comparable with human performance (man can deal with about 15 images per second). There is a direct relation between the complexity (and the effectiveness) of the algorithm employed and the time taken to obtain the result.

(b) In classical pattern recognition very large memory capacity and powerful and fast processors are available. A drawback when specifying robot systems is that they can only be equipped with a computer comparable in price with that of the articulated mechanical system itself. Microprocessors or a minicomputer of limited speed and of restricted memory are therefore used. It has been estimated that, to carry out complex (three-dimensional) image analysis in real time, a system capable of executing 100,000 operations per second[52] would be needed, using available form recognition algorithms and a classical information system.

(c) The interpretation of an image in ordinary pattern recognition is extended for off-line situations where man can exercise his judgement on the computed result. This is usually merely a proposition generated by a machine with no responsibility for its consequences which can be criticized before action is taken.

This is not true for a robot — it will make its decision and act on the basis of image analysis, irrespective of the consequences. Greater safety and confidence are the result of the recognition process required from the robot. For this reason, algorithmic recognition or decision processes have to be treated with care.

(d) Pattern recognition is particularly concerned with two-dimensional scenes (writing, images obtained from satellites etc) but the robot has to act in a three-dimensional universe. It manipulates objects which possess *volume* so the *relief* of the object is of importance.

(e) Often, the recognition methods used are based on specific qualities of the objects to be recognized. For example, two objects might be distinguished, one from the other, because one contains a hole. Such features might be used in practice, but the method employed must be sufficiently versatile to cope with a range of unforeseen features.

(f) The last essential difference concerns the extraction of information specifically required for the robot to carry out its job. This can be done by developing an algorithmic simplification and it is in this direction that work on robotic vision needs to be pursued.

So, plainly, the techniques developed in pattern recognition have to be considered afresh before integrating available technologies. Straightforward application of existing methods can only be used in specific cases. In general, more vigorous constraints in form recognition have to be applied to robots than in other applications.

6.4 Positioning of visual sensors

In deciding how to position visual sensors, assuming a fixed robot in a restricted space acting on specified objects, the nature, position and orientation in space of each object have to be taken into account. The object in question must obviously be 'seen' by the sensors, but, in addition, the characteristics and positioning of the sensors will have to be determined by the nature of the visual information required. The need to obtain a relief image is an important case in point and will be considered later. One assumes that two-dimensional (2D) sensors are available. A combination of two or more sensors, used to obtain depth information, is called a *K-2D sensor* (this can be used for a single, repositionable 2D sensor). Single sensors giving depth information are called *3D sensors*.

Figure 6.4 shows the positioning of sensors for a robot used to manipulate objects placed in front of it (whether in a fixed or mobile plane, or in a container). Use can be made of fixed cameras placed in position in the work area or mobile cameras integrated with the robot, either contained in its structure (if possible at its end effector: 'the eye in the hand') or on an ancillary mobile structure whose movements are actuated by those of the robot itself.

6.5 Sensors and depth reconstitution

Consider a three-dimensional object which needs to be reconstructed (eg to manufacture identical copies) from given 'photographs'. Two factors must be known:

1. The relationship between size of the photograph image and the size of the object ie the scale.
2. The shape of the surfaces of the object or the coordinates of each of the points which make up the surface. Photographs that show *all* the information required to reconstitute the surface are needed. There are two categories of objects:
   (a) Those whose surfaces contain hidden points, irrespective of the angles used when taking the photographs (Figure 6.3).

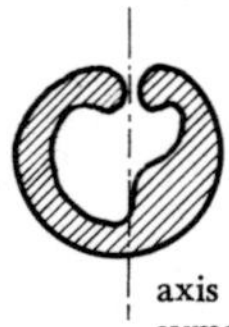

Figure 6.3. *Section of an object with axial symmetry which can never display some point on its surface (the camera is incapable of entering the object)*

   (b) Those for which all the points on their surface can be photographed.

A 2D sensor can be effectively created by using a 1D feeler (for example, a linear array of photodiodes) which scans the scene (with a single DOF) or by using a single sensor (0D) which can also scan the scene, with two DOF given enough angles of exposure. Among these are objects possessing convex (Figure 6.5) and concave surfaces.

It is possible to use only two photographs in dealing with a convex surface if the viewing angles are well chosen. If, on the other hand,

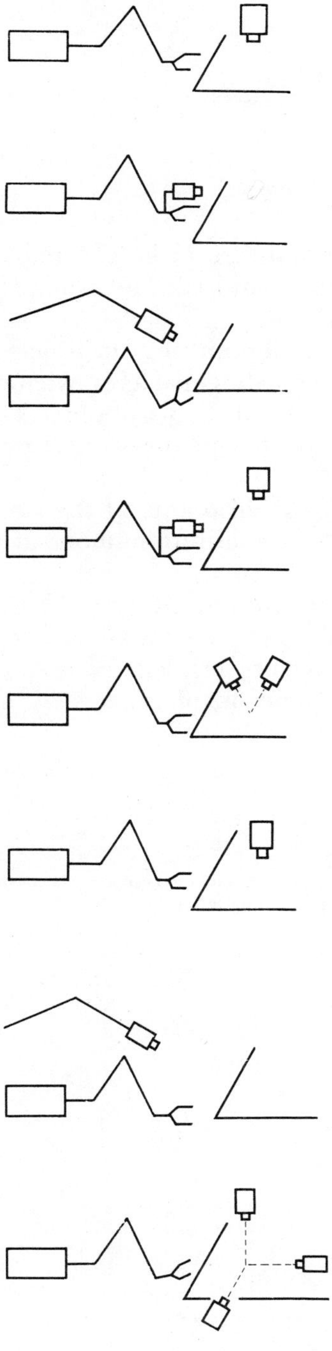

(a) 2D sensor fixed above the scene

(b) 2D mobile sensor fixed to the end effector. Sensors integral with the gripper are being researched

(c) 2D mobile sensor driven by an ancillary structure which is often Cartesian

(d) Combination of a 2D fixed and a 2D mobile sensor. The mobile sensor could be auxiliary

(e) Two fixed 2D sensors. The angle between them is optimized according to the task

(f) 3D fixed sensor

(g) 3D auxiliary sensor can be integral with the robot given appropriate technology

(h) Three 2D sensors giving projections on orthogonal axes

Figure 6.4. *Some positioning options for sensors on manipulators. These are not only a function of what needs to be seen, but also of the available technology*

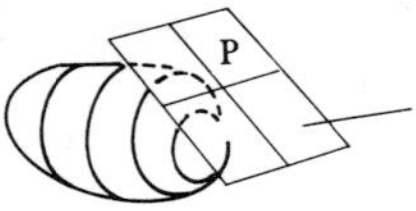

Figure 6.5. *Convex object*

fixed angles are used and all possible points are to be covered with all possible orientations, six photographs are required. The example of the cube is shown in Figure 6.6.

For a convex object 'floating' in space all points on the object can be seen, using six cameras situated on three orthogonal axes, whatever the orientation of the object. This is not true for concave objects. Fixing viewing angles cannot be used to ensure that pictures of all points of the surface will be taken.

To be certain of obtaining images from all points of the surface of concave object, one needs, theoretically, an infinite number of camera directions, covering solid angle 4π.

A robotic visual system, in practice, would, either (a) be able to cover an infinite number of camera angles (for a concave object) or (b) not be able to see certain parts of other objects. The consequence of (b) is extremely important in the recognition of objects by a robot

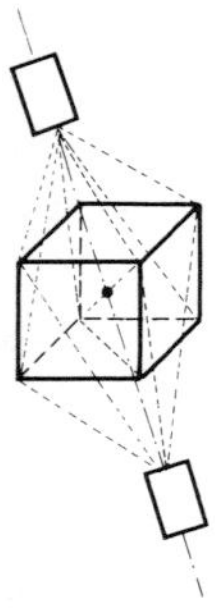

(a) Two well-sited cameras can cover the entire surface

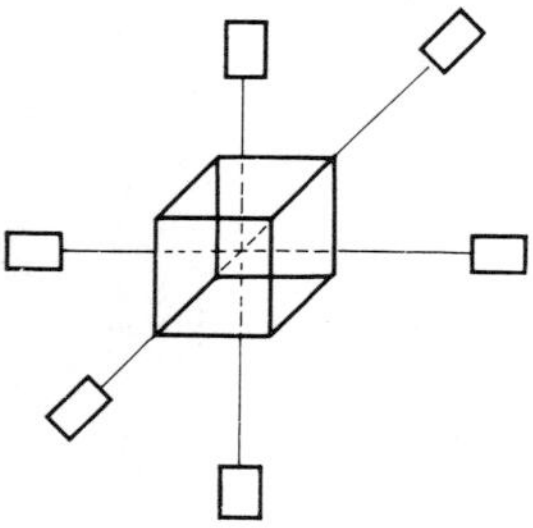

(b) Six viewing positions at most are needed to cover the cube

Figure 6.6.

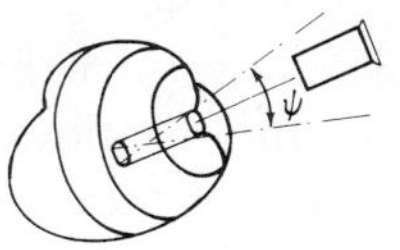

Figure 6.7. *Viewing a concave object*

equipped with vision. If, in the learning phase, the robot memory can be supplied with a complete description of a three-dimensional object (eg the coordinates of the group of points on its surface in the set of coordinates of the object), it becomes evident that the recognition phase must be based on incomplete evidence. Although mathematically such recognition can never be absolutely certain, as in man, one must allow for an element of guesswork.

To reconstitute relief we will have to go beyond a smooth image and obtain 'information' on the third dimension. This has been done using several techniques which are called (see above) 2D, K-2D and 3D.

6.5.1 DEPTH SENSING AND THE TWO-DIMENSIONAL METHOD

The 2D technique provides a smooth image from which information concerning the third dimension cannot be obtained. 2D may be used for a robot that handles objects if one (or the other) of the following conditions is fulfilled:

(a) The object has, effectively, only two dimensions, for example, the very flat parts resting on a horizontal plane, the axis of the camera being vertical.
(b) The problem is reduced, artificially, to two dimensions by pre-storing information on the third dimension in the robot.
(c) All possible images of the scene as seen from all possible angles are stored and an optimal viewing angle obtained.

Suppose the parts in Figure 6.8 are of interest. Telling the robot, before recognition, that part A has a height h_1 and part B a height h_2, a two-dimensional examination of faces F'_1 and F'_2 will be sufficient

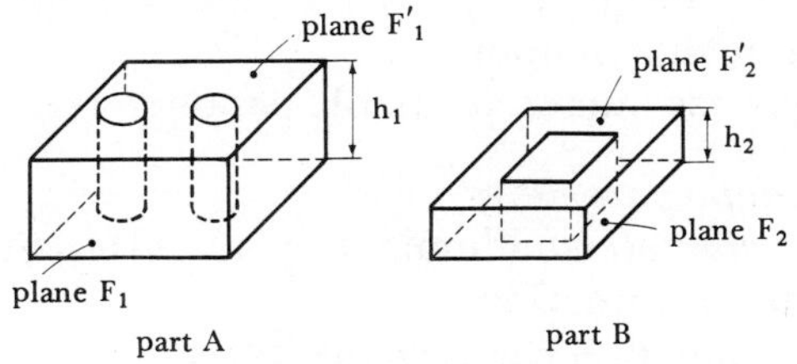

Figure 6.8. *2D recognition*

to provide complete knowledge of A and B. This technique of providing information *a priori* in the third dimension is often used.

When the object possesses several possible support faces, the 2D technique is based on the assumption that the number of possible objects is equal to the sum of the possible support faces. In Figure 6.8 that indicates 12 different objects (six possible support faces per object). Visual examination of the face opposite from the support face allows the robot to discover its position and its spatial orientation.

6.5.2 THE K-TWO-DIMENSIONAL METHOD

Using technique K-2D it is hoped to obtain depth information by taking several non-identical images. This technique finds its origin in human stereoscopic vision which, using two sensors (eyes) to view the scene from two different angles, creates a sensation of depth.

In fact, two images obtained by two fixed cameras need not provide sufficient information to produce an adequate image in relief. So, as in Figure 6.4, multiple images, using a number of fixed or mobile cameras, are obtained, using a range of angles. When using mobile cameras a strategy has to be worked out that takes into account known information. For pictures from different angles a fixed camera taking mobile objects can also be used, for example, objects which go past on a conveyor belt. Equally images using radiation at different wavelengths could be taken.

6.5.3 THE THREE-DIMENSIONAL TECHNIQUE

Technique 3D is used when depth of information is obtained not through comparison of images taken from different angles, but through the use of direct information on the third dimension. This can be obtained in several ways.

Consider a 2D image presented to a sensor. What could bring out component r(X, Y) of this image (section 6.2), the image being defined by g(X, Y)? To do this one needs to reconsider the nature of a 'complete' sensor. A transmitter illuminates the scene, which reflects light to a fixed receiver (assuming that optical radiation is used). To obtain 3D information in a 2D image, properties of the transmitter or properties of the scene must be drawn upon.

A source of light has numerous usable parameters:

1. *Geometric shape parameters.*
2. *Frequency parameters*: illumination at different wavelengths can provide useful information.
3. The incident radiation can be *polarized.*
4. The light used can be made optically *coherent.*

5. *Variable parameters:* the light can be modulated in various ways – spatially: directionally ie by scanning space using a bundle of rays, transverse modulation of the bundle transmission of pulsed light of (a) varied wavelength or (b) varied angle of polarization.

The reflected pencil of light will contain information related to the scene. It will contain components that reach the sensor by diffusion, diffraction (Fresnel and Fraunhofer), reflection and absorption. At the sensor, variations in g(X, Y) will contain components that are functions of (X, Y), functions of time or of other variables such as the wavelength of radiation.[53] Depth information will be affected by these other components in the received optical signal.

All of these factors have been considered in several studies of three-dimensional sensing.[54,55] The function r(X, Y) is not always recorded at the sensor g(X, Y) but sometimes at an auxiliary sensor.

6.6 Image sensors

These sensors are called *active* or *passive* according to whether or not they contain a light emitter. In fact, only 3D sensors are, or can be, active.

Image sensors are systems that convert an image formed by specific radiation (visible light, ultrasonic, X-rays etc) into an electric signal exploitable by an appropriate device.

The optical sensors used include television cameras, image converters and active systems (transmitters included). In the ultrasonic domain one also finds various echo-based imaging methods.

6.6.1 'TELEVISION' TYPE CAMERAS[56]

6.6.1.1 Image converter[57,58]

This traditional device comprises a vacuum tube containing a photocathode.[59] A system of electron lenses directs emitted electrons through the anode aperture, which results in an output from the multiplier proportional to the incident light. The region of interest on the photocathode can be selected by controlling the electron lenses, which thus creates a system with variable access. The image itself can also be scanned but if the scan is too fast, the time constant is too long to provide high resolution and low noise. This is a slow system, but one capable of high resolution, better suited for inspection in front of or behind the robot rather than as a continous monitor or 'eye' for the robot.

Nevertheless, such devices have been used to compare two images in real time,[60] using the fact that the detected current, which corresponds to a specific luminous point, is a function of:

(a) The luminosity of the point.
(b) The exposure time.

If the latter is modulated by a second image, correlation is possible.

6.6.1.2 Flying spot scanner[61]

This is a variant of the image converter used to analyse negatives of photographs that are placed against a phosphorescent screen. The light produced at a point is focused onto a photomultiplier. Thus the negative can be scanned point by point and excellent resolution is obtained. This system is used, for example, in the examination of negatives obtained from a bubble chamber, where such fine resolutions have been obtained, allowing a 17 bit code to be used.[62] From this it may be assumed that the system is quite slow but very precise and thus more suitable for inspection purposes rather than for handling.

6.6.1.3 Vidicon-type tubes[63]

The Vidicon tube in one form or another is used in all, or very nearly all, current television cameras. Unlike the image converter it possesses an electron gun, which scans at constant speed a photoconducting mosaic laid down on a transparent metallic film. When light passes through the film it hits the mosaic, changing its resistivity. The current in any element of the mosaic depends on its resistivity, which is determined by the incident luminous energy, and is also affected by the time interval between successive light pulses. To obtain a signal that depends only on the luminous energy, the time interval between pulses must be constant whatever the position on the mosaic. Access to the system must be strictly sequential, unlike the image converter that allows random access.

The nature of the mosaic used greatly affects performance. It has been possible to increase the output signal and the spectral range using the silicon Vidicon.[64] In this, the mosaic is replaced by a silicon layer with diodes diffused on one of its sides. Plumbicon too improves performance,[65] the diodes being made with lead monoxide.

Vidicon cameras are cheap, because of their extensive use, and are less cumbersome than the other devices available. Although they are fragile they are very reliable. This is why they are much used in robotics and are only now about to be replaced by solid-state devices and other recent developments.

6.6.2 CHARGE-COUPLED DEVICES[66,67]

These employ 'solid' sensors rather than fragile vacuum tube devices. In fact, they are equally sensitive to shock but have many weight and volume advantages and are compatible with digital systems.[68] They are therefore the sensors of the future, able to be integrated into the structure of the robots themselves, particularly into the end effector.

They are structured as arrays or matrices of photosensitive elements. These elements can be MOS devices or photodiodes:

1. *MOS devices:* The incident photons create electric charges in the silicon. These accumulate under the polarized grid of the MOS device, either at the silicon-oxide interface (*surface CLD*) or in the bulk of an implanted channel (*bulk CCD*) (Figures 6.9 and 6.10). The MOS devices are isolated, one from the other, by an auxiliary electrode or by an end channel. The photons can enter from the electrode side (front exposure) or from the side of the reduced substrate (back exposure).

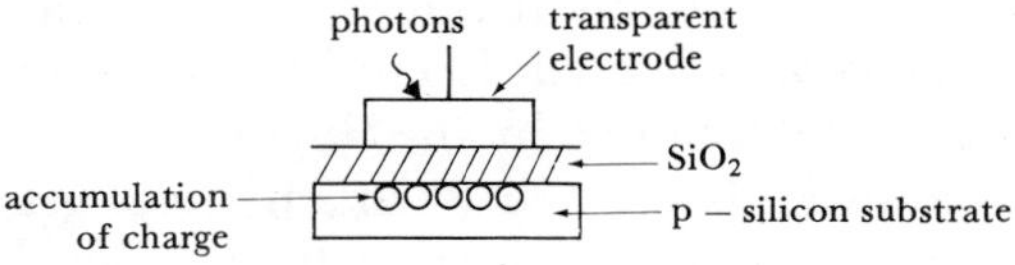

Figure 6.9. *Surface CCD*

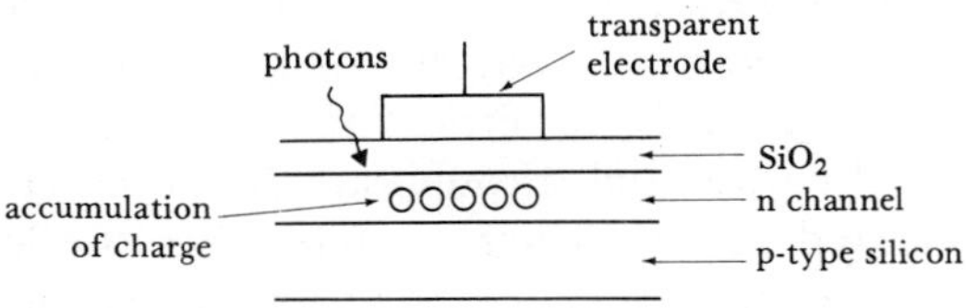

Figure 6.10. *Bulk CCD*

2. *Photodiodes:* The photosensitive element is a reverse-biased photodiode. The charges created by the photons are stored under the electrode of an MOS-like structure (Figure 6.11). The CCDs comprise, apart from the photosensors, a reading register and appropriate output amplifiers.

The charges under the photosensitive element are transferred to a register which transfers the useful signal step by step towards the output by applying variable voltages to the electrodes. The charges accumulated in the register are read by a charge amplifier which measures the voltage variations across the terminals of a capacitor. Between two successive readings, the capacitor is recharged with a reference voltage.

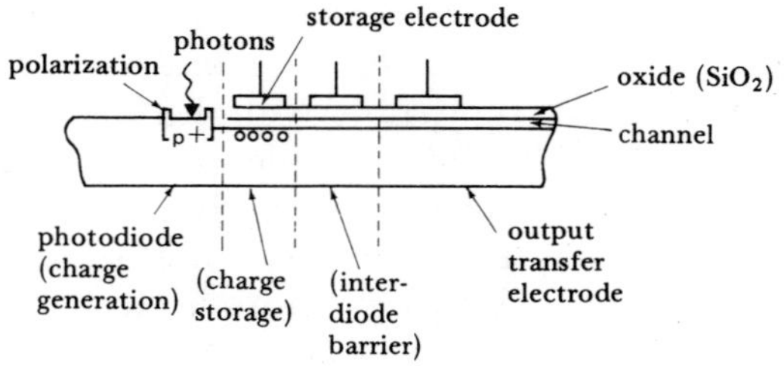

Figure 6.11. *Photodiode structure*

6.6.3 IMAGE CONVERTERS[69,70]

These are also called *coherent* or *incoherent* image converters, or *spatial light modulators* and are based on the use of images in real time.

An image is created using incoherent light on an active crystal. The crystal is illuminated by a coherent bundle of light rays. The coherent light bundle, emerging or reflected, is modulated by the local intensity of the image. In a sensor developed by the Hughes Research Laboratory[71] (Figure 6.12), a sandwich is formed from a plate of glass, a transparent conducting electrode, a photoconducting layer of cadmium sulphide, a mirror, a liquid crystal, another electrode and another plate of glass. An alternating voltage is applied between the two electrodes.

The intensity distribution of the incident light (right-hand side in Figure 6.12) is converted into a pattern of photoelectron-hole pairs in the cadmium sulphate photoconductor. The alternating voltage promotes a redistribution of the potential throughout the liquid crystal.

A linearly polarized pencil of light enters from the left-hand side of the system (Figure 6.12). It is reflected by the dielectric mirror, crosses

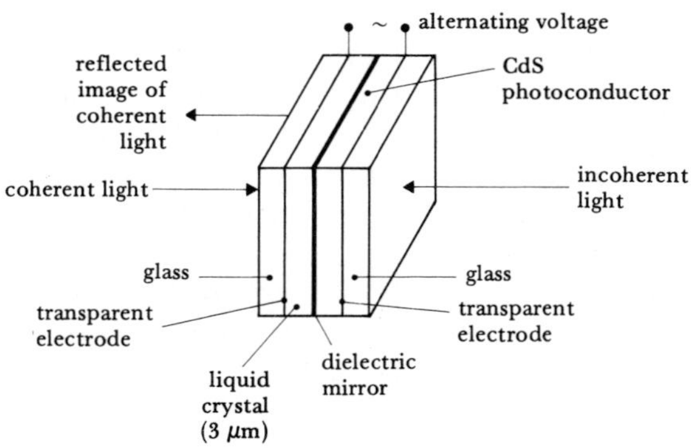

Figure 6.12. *Liquid crystal transducer developed by Hughes Research Laboratory*

the liquid crystal for a second time and leaves the system elliptically polarized. After it has left the converter an optical analyser, situated in the path of the pencil of light, only lets the perpendicularly polarized component pass in the original direction of polarization of the reading bundle. After passing through the analyser, the spatial distribution of intensity of light in the reading bundle is approximately proportional to the spatial intensity distribution of the incident image. In the prototypes that have been tested the authors obtained a resolution of 25 μm for a surface image of 25 mm^2. The contrast ratio was 100 : 1 and the reading time as little as 30 milliseconds.[70]

In addition to this type of optical converter there are also electronically addressed transducers (Figure 6.13).[72] Their future lies in the processing of information, which is much faster than processing by software.

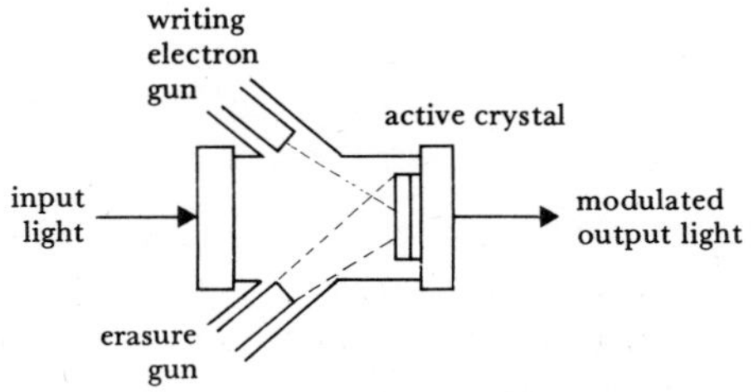

Figure 6.13. *Electronically addressed transducer*

6.6.4 ACTIVE OPTICAL IMAGE SENSORS

By measuring the reflected light from the scene illuminated with a pencil of rays, distances may be measured to provide relief information. Using our terminology, these are 3D sensors.

The first technique consists of sending a bundle of known cross-sectional shape (flat or cruciform)* and then recording the image of the reflected pencil of light using a television camera or a solid-state camera.[73, 74] A pencil of light can also be used, with the direction of the reflected light being found.

A second technique consists of scanning the scene to be recorded using a laser. Either the transit time of a reflected pulse of light[75] or the phase change of the amplitude modulated signal reflected by an incident pencil of light can be measured[76] as shown in Figure 6.14(b).[77] Opto-acoustic detectors can be used instead of rotating mirrors.

It has already been noted that multiple devices may be used to obtain relief information.

* Some research workers, such as Armstrong and Cadieux[78] in Montreal projected a grid formed of black and white squares randomly shaped and positioned. The deformation of this grid (Figure 6.14a) is observed.

Figure 6.14(a). *Random grid*

Stereovision has not been mentioned here – measurement by two cameras which see the scene from two different angles. Triangulation using stereovision presents three kinds of difficulties:

(a) Since the differential of the two images is made, the relief data derived are very sensitive to noise.
(b) Correspondence between the points on two images of the same scene is not easy to establish.[79]
(c) Measurement of distant points is less precise than that of near points.

6.6.5 ULTRASONIC IMAGING DEVICES[80]

Ultrasonic devices use frequencies of several megahertz in 'classical' image makers or several gigahertz in ultrasonic microscopes.[81] Using such methods, images of internal cross sections even of opaque solids can be obtained. So, ultrasonic image makers are 3D sensors.

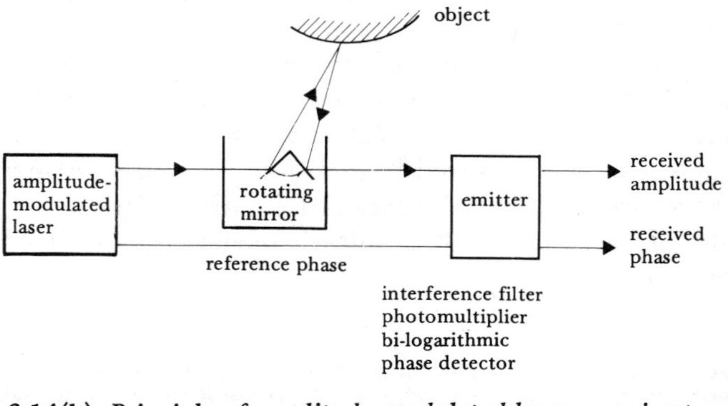

Figure 6.14(b). *Principle of amplitude-modulated laser scanning transducer*

Their fields of special application are in observing living tissues and in non-destructive industrial inspection. In robotics, the application would be in farm-product inspection.

Such ultrasonic methods all function according to the basic principle of radar. They are generally classified according to the method of representing information used. Echography (or sonar techniques) can be classified as type A, M or TM, B and C.

1. Echography A: An ultrasonic signal is injected from a source into the body to be examined, the reflected signals originating at interfaces between media of different acoustic impedence. The echoes are picked up by the same device and their amplitude is a function of the reflectance of each interface. The distance between source and interface translates as elapsed time.

Uni-dimensional echography (ie method A) is being replaced by bi-dimensional echography (method B).

Method M of TM (time motion) is a variant of method A but used with mobile objects.

2. Echography C:[82] In method C the image of a plane situated at a fixed distance from the sounding-line and normal to the transmitting pencil of light is obtained. This (or a focusing aperture) moves on a surface (Figure 6.15) and only echoes reflected from a predetermined distance are selected by an electronic gate system. By varying this distance, readings from successive planes of the object are obtained, these providing depth information. Depending on the version used, the scanning may be mechanical or electronic, the lenses acoustic, electro-acoustic etc. Up to about 100 images per second can be obtained.

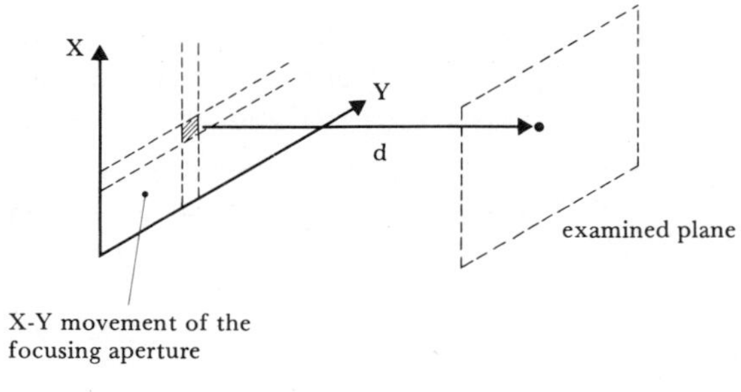

Figure 6.15. *C-type echography*

Although little or not at all used at present in robotics, perhaps because of the difficulty of image processing, its size and its complication, the ultrasonic image-maker possesses, by its nature, a very important advantage over optical sensors, namely its independence of lighting, a major problem for robots that can see, as will be seen later. However, its effectiveness at a distance remains poor.

6.6.6 DEFECTS IN OPTICAL SENSORS (TELEVISION TUBES AND DTC)

These are not serious faults and can even be beneficial. CCDs are generally linear. Silicon Vidicons are almost linear and:

$$I = KE^{\gamma}$$

where I is the current; K is constant; E is luminous intensity and γ is of value 0.95 to 1.

Logarithmic sensitivity allows a computer to distinguish more grey levels in the subject than does a linear scale.[83] Electronic systems that correct the intensity scale can be constructed. It depends on what is being sought in the image. It must not be forgotten that in the majority of industrial applications the image will be *linearized* through thresholding. The choice of threshold is more important than the linearization of intensity which is only important in sophisticated systems related to artificial intelligence.

6.6.6.1 Lag

This is the effect of permanence of an image when the object or the camera suddenly moves. This fault exists in Vidicon cameras and can have a duration of 90 milliseconds. It does not exist in CCDs.

It is a fault which becomes a handicap if the image acquisition has a duration comparable with the lag time. It is one of the reasons why research workers chose CCDs, especially when dealing with moving objects, for example on a conveyor belt.

6.6.6.2 Blooming

When a sensitive surface is strongly illuminated at a point the area that emits light as a result spreads beyond the original point. A way of reducing such blooming consists of making the sensor response above the threshold non-linear. To do this special guns (ACT or anti-comet tail) are used[84] in Vidicon tubes. An implantation in the cathode substrate has also been known to help.[64]

6.6.6.3 Spectral response

This stretches from 0.4 μm or 0.8 μm to 1 μm according to the tube in question. CCDs are often sensitive to infrared and this should be taken into account when filtering emitted radiation.

6.6.6.4 Parabolic distortion

This only exists in tubes that use an electron gun for searching. It is an intensity variation between the centre and the edges of the sensitive surface. In other words, if the sensor's surface is uniformly illuminated, the current detected is greater at the centre than at the edge. Such distortion arises from secondary emission, which depends on the angle of incidence of the electron beam. It can be modelled and can be corrected electronically.

6.6.6.5 Image noise

This is essentially created in three ways:

(*i*) *Background noise* due to sensitivity fluctuation at a point on the sensor. This is due to fluctuations in the number of electrons in the scanning bundle or in the number of hole-electron pairs created by the passage of the bundle.
(*ii*) *Thermal noise* produced in the resistances in the circuits.
(*iii*) *Noise* originating in the *amplifiers*.

Image noise is seen in the image as randomly placed departures resulting in problems of interpretation, as one must decide at each instant whether an intensity is due to the object in view or noise.

In solid-state CPD sensors, noise is seen as sensitivity differences between photosensors or even in a whole line in respect to the other lines. A typical CCD matrix of 40,000 points, for example, may have 10 faults (10 photosensors which do not work). Such a product would be one-half as expensive as a faultless matrix. On the other hand, when such faults are localized they can be modelled in the processing.

Background noise is also present and is often treated with filters.

6.6.6.6 Conclusions

It has been shown that in television tubes or CCDs, the principal faults can be corrected. Blooming and background noise cannot be eradicated, but only attenuated; for blooming, the scene illumination can be reduced to improve matters. Background noise remains, which alone presents difficulties.

Conclusions

In this chapter the problem of vision in fixed robots has been considered and the principal kind of practical sensors described. As expected, no sensor is perfect but adequate images can be obtained in practice.

It should not be forgotten that the resulting images are distorted and noisy representations of the original scene, from which one must reconstruct what one is looking for, bearing in mind, as far as possible, what is known about it already.

Chapter 7

Remote sensing: artificial vision—representation of visual information

In previous chapters it has been shown that an image consists of analog data as the output of a large number of sensors. How are these data to be handled?

First, the view might be taken that processing these data will be particular to the required task of a specific robot. Secondly, there is the almost opposite view that, as in human vision, it may be possible to extract universal representations from this jumble of data. It is only the use of such representation that becomes specific for specific tasks. Thirdly, and whatever the case, it must be understood that the processing system must take decisions within milliseconds or, at worst, tens of milliseconds. The fact that the processor can not only interpret these data but also control their acquisition should not be forgotten.

Given these three points universality should not be confused with optimality, in the sense that good performance in image interpretations must focus on attaching meaning to images. To do this, optical, analog or digital aids could be used.

7.1 Representation of an image

Starting with the function g(x, y), which represents the intensity at each point of the sensor's surface (in a 2D sensor), representation consists of processing g(x, y). This will lead to a *new presentation* of the image possessing two characteristics:

1. The information needed (for further action ie *pertinent* information must be included).
2. The data must be simplified so as to ease the task of subsequent decision-making processes.

This can lead to:

(a) Partial evaluation or extraction of the information required.
(b) Dilution or loss of some of the information contained in g(x, y).

This filtering and structuring phase leads to the acquisition of a *characteristic image* or a set of *characteristic features*. This representation often occurs in two phases. The first is *image enhancement* where details will have disappeared, with others thrown into relief. The second is *coding* the image which can be interpreted by the computer. Coding represents, on a kind of data base, the key features of the image. These are generally available in *geometrical*, *functional* or *topological* form from the enhanced image.

The process of transforming the image can vary with the type of sensors and class of application. For example, in optical processing of images, the sensors can supply directly a functional representation in the form of a Fourier-transformed image of the scene.

7.2 Image enhancement

This comprises two aspects:

1. The suspension of defects originating in the sensor ie differences between the image presented to the sensor input and that derived from the recording of the electric signal that it delivers.
2. The reinforcement of elements that enrich the required information and suppression of other, unwanted, elements. Enrichment and suppression can be treated simultaneously in certain cases and separately in others.

7.2.1 ENHANCEMENT AT THE SENSOR

Having decided that most defects in the sensor may be dealt with, it should be noted that two may be handled at the analog level. These are background noise and linearity of the signal as a function of the luminous intensity. If background noise is eliminated, the question of linearity still remains. This is now inescapably related to a consideration of the spatial nature of the image. It is not necessarily the level of luminous intensity but *differences* in intensity across the image that contain such information. Such variation leads to contrast lines or outlines, and the shape and the distribution of these outlines contain the information required. Consequently, improvement of linearity at the image sensor is needed to extract contrast lines with some reliability. This can be done by modifying the characteristics of the sensor amplifiers. Filter techniques can also be used, in order to reinforce certain elements in the image.

7.2.2 ANALOG FILTERING

In the representation process there are both analog and digital images respectively. Filter techniques can be applied to one or both. They can also be *implanted* by digital or analog (eg optical) means, depending on needs. Analog filters that may be applied to the analog image derived from the sensor will be considered first.

7.2.2.1 *Spatial smoothing*[85]

This is used to counter noise. Consider a window of arbitrary shape on the image, of surface S_w, centred on the point x_i, y_j. $g(x_i, y_j)$ replaced by:

$$g'(x_i, y_j) = 1/S_w \iint_{w(x,y)} g(u, v)dudv \qquad (7\text{-}1)$$

Its dimension can be similarly treated, so that $g(x_i)$ is replaced by:

$$g'(x_i) = 1/w \int_{x_i - w/2}^{x_i + w/2} g(u)du \qquad (7\text{-}2)$$

The principle consists of replacing the intensity of a point with the average of the intensities of the adjacent points. As shown in Figure 7.1, an effect of softening or defocusing is obtained that can make certain details disappear. Note also that the smaller S_w is, the more the initial image is modified. In each case it is a matter of finding the optimal area that diminishes the noise without losing pertinent details or outlines.

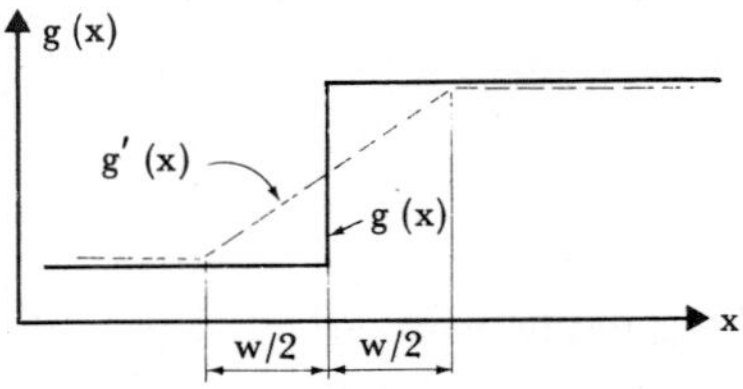

Figure 7.1. *Effect of spatial analog smoothing*

7.2.2.2 *Spatial filtering*

Consider an image represented on an infinite plane by $g(x, y)$. Its Fourier transform is written:

$$G(f_x, f_y) = \iint_{-\infty}^{+\infty} g(x, y) \cdot \exp[-2\pi i(f_x x + f_y y)]\, dxdy \qquad (7\text{-}3)$$

to which the inverse transform corresponds:

$$g(x, y) = \iint_{-\infty}^{+\infty} G(f_x, f_y) \cdot \exp[2\pi i(f_x x + f_y y)]\, df_x df_y \qquad (7\text{-}4)$$

f_x and f_y are called *spatial frequencies* analogously with the frequencies x in the temporary Fourier transform. The more changes, the higher the spatial frequencies involved.[86] In other words, contrast lines or outlines are going to lead to high values of f_x and f_y and weak contrast zones to low values of f_x and f_y.

This leads to the use of high-pass or low-pass filters to improve the image g(x, y) after obtaining its Fourier transform $G(f_x, f_y)$. $H(f_x, f_y)$ is the transference function of a linear, spatial filter applied to $G(f_x, f_y)$. The Fourier transform of the filter output is then:

$$G_s(f_x, f_y) = G(f_x, f_y) \cdot H(f_x, f_y) \quad (7\text{-}5)$$

and the image after filtering will be represented by:

$$g_s(x, y) = F^{-1}[G_s(f_x, f_y)] \quad (7\text{-}6)$$

$$= F^{-1}[G(f_x, f_y) \cdot H(f_x, f_y)] \quad (7\text{-}7)$$

$$= g(x, y) * h(x, y) \quad (7\text{-}8)$$

with:

$$h(x, y) = F^{-1}[H(f_x, f_y)] \quad (7\text{-}9)$$

The asterisk represents the convolution product and h(x, y) is the impulse response of the filter. A filter of the form:

$$H(f_x, f_y) = 1.5 - (\cos \pi f_x \cdot \cos \pi f_y)^4 \quad (7\text{-}10)$$

is programmable. It is a high-pass filter which alternates low frequencies and favours high frequencies, as is evident from the calculation. The synthesis of optical filters is more complex.[70]

7.2.2.3 Spatial differentiation

This consists of replacing each point of g(x, y) with its gradient g*(x, y) in absolute value:

$$g^*(x, y) = |\nabla g(x, y)| = [(\partial g/\partial x)^2 + (\partial g/\partial y)^2]^{1/2} \quad (7\text{-}11)$$

Consider a one-dimensional image in which g(x) has the form indicated in Figure 7.2. It can be seen that differentiation brings out the outlines and eliminates the zones where g(x) is uniform. This is why differentiation is often used, either at the analog stage (in optical processing) or at the numerical stage.

Note that the use of spatial differentiation and the low-pass spatial filter have opposite results to those obtained with spatial smoothing and the high-pass spatial filter. The order of application is very important. In fact, differentiation applied to background noise simplifies it considerably; with weak outlines noise elimination becomes difficult if carried out at the smoothing stage. The smoothing or 'cleaning' stage is always carried out first.

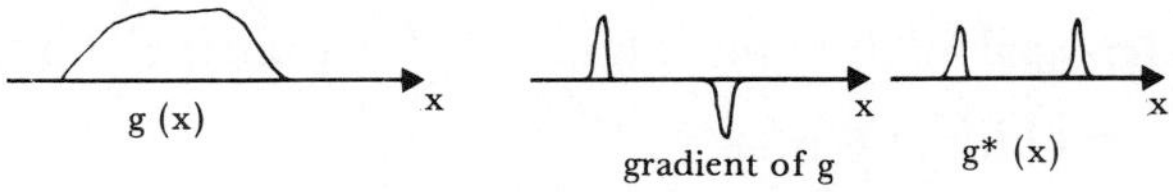

Figure 7.2. *Signal differentiation*

7.2.3 DIGITIZATION OF THE ANALOG IMAGE

The computer can only operate on a digital image. g(x, y) must therefore be digitized. Note that this depends on two elements:

1. The sampling of the image as n $\times$ m points forming n lines and m columns.
2. Intensity g which will be coded in K bits representing 2^K grey levels.

Three elements must be taken into account to determine sampling intervals both in surface and in time:

(a) The total time taken by the computer to digitize and acquire the image.
(b) The amount of data obtained that must be stored and subsequently processed.
(c) The resolution needed from the image, in amplitude as well as in space.

Let us examine this last point first by beginning with analog image g(x, y) and by considering which *spatial* sampling could retain information, considering, for simplicity's sake, a one-dimensional image g(x, y). As in equations (7-3) and (7-4) the Fourier transform can be written:

$$G(f_x) = F[g(x)] = \int_{x=-\infty}^{x=+\infty} g(x) \cdot \exp(-2\pi i f_x x)dx \qquad (7\text{-}12)$$

$$g(x) = F[G(f_x)] = \int_{f_x=-\infty}^{f_x=+\infty} G(f_x) \cdot \exp(2\pi i f_x x)dx \qquad (7\text{-}13)$$

Suppose f_x, the spatial frequency, is restricted to a range of frequencies:

$$-f_x^* < f_x < f_x^*$$

In the Fourier series:

$$G(f_x) = \sum_{m=-\alpha}^{+\alpha} C_m \cdot \exp(-2\pi i f_x m/2f_x^*) \qquad (7\text{-}14)$$

with:

$$C_m = 1/2f_x^* \int_{-f_x^*}^{+f_x^*} G(f_x) \exp(2\pi i f_x m/2f_x^*)df_x \qquad (7\text{-}15)$$

$G(f_x)$ is also zero at the exterior of the strip $(-f_x^* + f_x^*)$; equation (7-15) is therefore identical with equation (7-13) for points $x = m/2f_x^*$. And:

$$C_m = (1/2f_x^*)\, g(m/2f_x^*) \tag{7-16}$$

Substituting in equation (7-4):

$$G(f_x) = 1/2f_x^* \sum_{m=-\alpha}^{+\alpha} g(m/2f_x^*) \exp(-2\pi i f_x m/2f_x^*) \tag{7-17}$$

$$\text{for } -f_x^* \leqslant f_x \leqslant f_x^*$$

$$G(f_x) = 0 \quad \text{for } -f_x^* > f_x > f_x^*$$

By taking the inverse transform of equation (7-17), $g(x)$ is obtained in the form:

$$g(x) = \sum_{m=-\alpha}^{+\alpha} g(m/2f_x^*) \frac{\sin 2f_x^*(x - m/2f_x^*)}{2f_x^*(x - m/2f_x^*)} \tag{7-18}$$

Because f_x^* is known to be the highest spatial frequency involved, then, applying Shannon's theorem on sampling, $g(x)$ can be reconstituted with the values deduced for distance intervals of $1/2f_x^*$.

A similar calculation may be made in two dimensions:

$$g(x, y) = \sum_{m=-\alpha}^{+\alpha} \sum_{n=-\alpha}^{+\alpha} g(m/2f_x^*, n/2f_y^*) \sin \gamma_x/\gamma_x \cdot \sin \gamma_y/\gamma_y \tag{7-19}$$

with:

$$\gamma_x = 2f_x^*(x - m/2f_x^*); \quad \gamma_y = 2f_y^*(y - m/2f_y^*)$$

In principle, each image contains a spatial frequency band in x and y, leading to ideal sampling. In practice, for technological reasons, spatial sampling cannot be adapted and because the spatial frequencies involved are not measured, the following procedure is carried out.

A total *number of points* is chosen to represent the image as opposed to a sampling step. With Vidicon cameras the choice is open, because it is sufficient to sample the video signal at a chosen frequency. If 1000 × 1000 points are chosen 'good' *spatial* definition is said to be obtained. Deterioration occurs when points are taken that are closer than $(1/2f_x^*, 1/2f_y^*)$. Solid-state cameras, however, have already been sampled spatially in construction; the only choice is to select fewer points than the sensor offers. This runs counter to Shannon's theorem.

Spatial sampling in effect introduces a *low-pass filter* whose results have been noted before.

In the coding of the amplitude of $g(x, y)$, precision is the result of the number of bits. Coding can be carried out, for example, in 8 bits

(256 levels of grey) or in any number of binary units. However, there is a limit above which extra binary units do not affect noise present in g(x, y), because of sensor limitations.

Suppose the digitization was carried out:

(a) According to Shannon's theorem.
(b) By coding in a number of bits that matches the precision of the analog sensor. The digital image available contains all the information present in the analog image. The latter could be reconstructed by starting the former.

Here is encountered a fundamental question which lies at the origin of numerous controversies and which separates, in the author's opinion, 'classical' recognition of pattern from the recognition of pattern applied to robots. Are conditions (a) and (b) above *necessary* for robots? Are they *useful*?

The digital image will be processed subsequently by the computer and the pertinent information contained in it ie that information which allows the robot to decide its behaviour will be considered. Pertinent information must not be confused with analog image. Not taking into account Shannon's theorem or the analog precision of the sensor, double filtering can be achieved. This can easily provide a numerical image in which all information pertinent to the problem is preserved. Therefore, filtering through digitization is highly recommended, for it reduces the number of sampled points and the bits per point (pixels), allowing information to emerge from background noise.

This, of course, supposes that one has a knowledge of required levels. They are divided into three categories:

1. *Logical* levels that correspond to a presence-absence procedure. For example, how many objects are present in the field of the camera? Knowing that there is no background noise and that the objects do not touch one another, it suffices to distinguish three clusters of points for which g is greater than a specified threshold to affirm that there are three objects. Shannon's theorem can be largely transgressed and the coding in 1 bit is sufficient.
2. *Topological* levels that correspond to relative positions of various regions in the image. Having objects A, B and C, is C between A and B, or to the left of A or B etc? There again, this can be done with an image which is, in quality, inferior to the analog image.
3. Levels based on *measurements*: counting points, calculation of surfaces, calculation of Fourier transforms etc. This depends on the precision needed: to distinguish a large object from a small one, precision can be very poor. It must be better if the objects are very close to each other. Levels based on measurement often activate detection of outlines: the finer the image, the better the outlines.

There is, then, an *optimal* quantification (see Wigglesworth[87]) of pertinent information. It seems that no systematic study has been made of this subject and that the problem has been considered as follows:

(a) Sampling is carried out with the shortest period possible and with a maximum number of grey levels so as not to lose any information. Then, the memory, digitization and processing of the image rapidly become incompatible with the use of small computers and therefore with robots with respect to cost and real time operation.
(b) Robots should not be used for a single task but for a category of tasks. To pass from one task to another, the processing algorithms of images will vary and will therefore require variable sampling. It is desirable to have precision equal to the maximum precision demanded by the most exacting algorithm, but precision must remain compatible with the constraints of real time.
(c) Several problems may be stored by the use of image comparisons. In these conditions, defocusing created by quantification is no longer a problem, up to a certain limit.

Taking this into account, images containing 10^4 points (matrices of order 100 X 100 or 200 X 200) discretized in grey levels (from 2 to 16) can be sufficient to resolve many problems connected with robots.

7.2.4 DIGITAL FILTERING

7.2.4.1 Spatial smoothing

Processes for an analog image are, of course, also applicable to a digital image. In principle, a given treatment can be applied either to one or the other, but not to both.

Spatial smoothing, as discussed in section 7.2.2.1, is applied by using a discretization of equation (7-1) and by considering a rectangular window whose sides are parallel with the lines and columns of the image matrix (Figure 7.3). If the rectangle has 2b + 1 points as a base and 2h + 1 points as the height, g(i, j) is replaced by:

$$g'(i, j) = [1/(2b+1)(2h+1)] \sum_{-b \leq m \leq b} \sum_{-h \leq n \leq h} g(i+m, j+n) \qquad \textbf{(7-20)}$$

The defocusing effect is then obtained.

A specific case presents itself when intensity g has only been coded in a single binary unit (the point is black or white). Functional regularization of equation (7-20) can be with advantage (ie to obtain the greatest speed) replaced by logical regularization. Figure 7.4 shows an example taken from the work of Duda and Hart.[86] Point e is left 'white' (equal to 1) if at least one of its adjacent points is also white. It is changed from black to white if all its adjacent points are white.

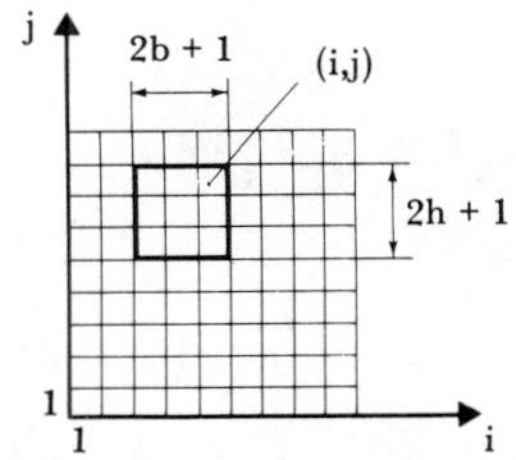

Figure 7.3. *Notations for spatial smoothing*

The expression then obtained is:

$$e' = \bar{e}(abcdefghi) + e(a + b + c + d + e + f + g + h + i) \quad \textbf{(7-21)}$$

In this expression the usual logical notation is used:

$$ab \equiv \text{a and b}$$
$$a + b = \text{a or b}$$
$$\bar{e} \equiv \text{inverse of e}$$

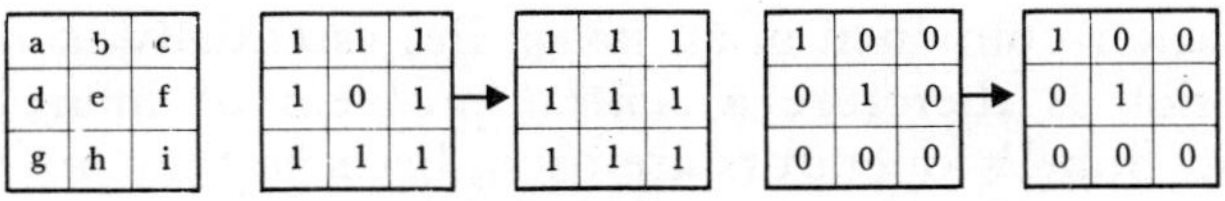

Figure 7.4. *Example of logical smoothing*

7.2.4.2 Spatial differentation

This is a procedure which throws the contrast into relief ie brings out the outlines. It is therefore used a great deal, for it is mainly outlines that are used in further processing. The two best known, most used, and simplest algorithms are those of Roberts and Sobel.

Roberts' Operator:[88] If four adjacent points (i, j), (i, j + 1), (i + 1, j), (i + 1, j + 1) are considered, an approximate gradient at point (i, j) using the function is:

$$R(i, j) = \{[g(i, j) - g(i + 1, j + 1)]^2 + [g(i, j + 1) - g(i + 1, j)]^2\}^{1/2} \quad \textbf{(7-22)}$$

or, again with the next function:

$$F(i, j) = |g(i, j) - g(i + 1, j + 1)| + |g(i, j + 1) - g(i + 1, j)| \quad \textbf{(7-23)}$$

is obtained.

Sobel's operator: This is applied to a window of 3 X 3 as shown in Figure 7.4. The gradient at point e is given by:

$$S = |S_x| + |S_y| \quad \textbf{(7-24)}$$

or by:

$$S = [S_x^2 + S_y^2]^{1/2} \tag{7-25}$$

with:

$$S_x = (c + 2f + i) - (a + 2d + g) \tag{7-26}$$

$$S_y = (g + 2h + i) - (a + 2b + c) \tag{7-27}$$

These two operators have the advantage of being quick and effective. They do not provide the real value of the gradient, but do provide an image on which calculations concerning the outline can be made.

7.3 Image coding

In the preceding steps the image has been pre-processed and put into a form that facilitates evaluation of its characteristic features.

7.3.1 DETECTION OF THE OUTLINES

The information contained in an image lies essentially in the outlines. Their detection is therefore a *central* problem of image processing. Roberts' and Sobel's operators are used, because they are very simple and adequate for robotic visual systems. Research workers have suggested many other, often more sophisticated (but cumbersome), operators without having obtained results that were totally satisfactory. An outline presents a double problem:

(a) Its detection.
(b) Its closure because it demarcates regions that have specific qualities.

A few methods that can be applied to a more or less filtered digital image, which can be of interest in robotics, will now be considered.

7.3.1.1 Thresholding

This is a binarization operation where value 1 is allocated to point x_i, y_j if $g(x_i, y_j) \geqslant g_{threshold}$ and value 0 if $g(x_i, y_j) \leqslant g_{threshold}$. Such a thresholding operation is always used at one stage or other of image processing, because the representation of outlines is either made in black on a white background or the other way round.

Choosing a threshold is of prime importance and the following procedures are used with varying degrees of success:

(a) Choice of a *fixed threshold.* For this procedure to be satisfactory, lighting conditions need to be strictly controlled.

(b) Use of the *same threshold* for all the points on the image, requiring a criterion for adaption to take place. One can, for example, envisage measuring the level of ambient light to control the threshold. Another method consists of making a histogram of the luminous intensity of all the points of the image and taking as threshold the abscissa in the 'valley' of the histogram if it is bi-modal,[89] or some other feature of this histogram.

(c) Use of an *adaptative threshold* for each point of the image. This solution could be of use provided the threshold criterion can be properly established. The criteria can be directly applied to the analog image, as Belan[90] and Rives[68] have done, for example. A value of 1 or 0 is assigned to a point of the image by comparing the video signal corresponding to this point with a signal c resulting from detection of the preceding modulation peaks in the video signal. The principle is demonstrated in Figure 7.5. The

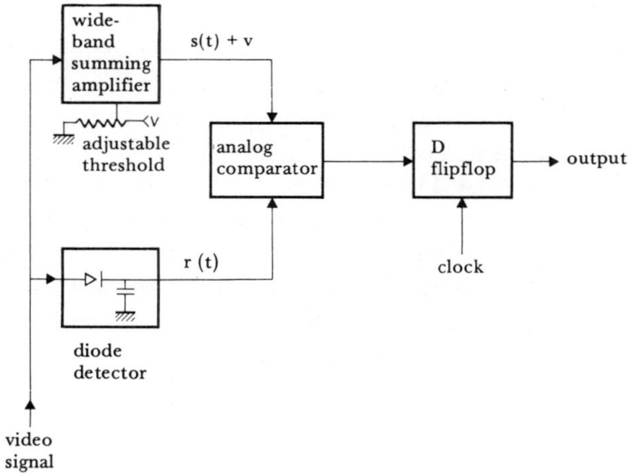

Figure 7.5. *Adaptive analog thresholding*

system comprises:

- An integrating amplifier, which defines a threshold for the comparator.
- A peak detector supplying dynamic reference (resulting from the value of capacity C).
- An analog comparator delivering the 0s and 1s for each point of the image.

The adjustable threshold can be made a function of the average illumination. A procedure of adaptive thresholding on the digital image can also be established. Chow and Kaneko,[91] quoted in Basseville,[89] chose the threshold in the following stages:

1. Dissecting the image into small, non-disjointed regions.
2. Histogramming the grey level in each region.
3. Selecting histograms of wide variance.
4. For each histogram, estimating the laws governing grey levels and the interaction coefficient using the method of least squares.
5. Testing the bi-modal characteristic of these estimated laws.
6. For each accepted bi-modal law, calculating the threshold, using maximum probability.
7. Interpolation to obtain a threshold for each of the regions, then for each of the points.
8. Making a binary decision for each point, starting with this threshold.

This example is instructive, but although it undoubtedly gives good results, it is much too complex and requires calculation times incompatible with the constraints.

7.3.1.2 Tracking outlines

Operators such as those of Roberts and Sobel often produce thick outlines (of breadth embracing several points). After thresholding it is a matter of reducing the width of, and regulating, these outlines (irregularity often arises from discretization).

The technique used by Dessimoz *et al.*[92] deserves to be mentioned. Two pointers trace two outlines, one the interior and the other the exterior. They move in such a way as to minimize their separation. The clearest outline is situated at an equal distance from the two pointers.

Other methods include the description of the lines by chain-coding (Freeman[93]) but these will not be considered here.

7.3.1.3 Other methods of extracting outlines

Processing images as described in the literature aims at optimal results in image detection, rather than aiming at the specific needs of the robot system. In robotics, the consideration of real time requires that simple methods be employed. The information obtained should be specific, and possibly capable of being reconstructed by the robot for its own purposes.[94]

Some mathematical methods that have been developed but that are of little practical application are:

- Use of a probability relation.[95]
- Use of gradient techniques followed by smoothing and thinning of the outlines.[96]
- Use of the Laplacian with iterative filtering.[97]
- Use of local comparison of averages taken in variable neighbourhoods.[98]

– Use of texture.[99, 100]
– Hueckel's operator.[101–103]
– Comparison with profile types.[104]

Heukel's operator: This well known method is used in five stages:

(a) On a surface D of the image a function S is sought which will define a value b from one side of a straight line 'to be found' and a value b + d from the other side. The equation of the straight line is written:

$$cx + sy = r \tag{7-28}$$

So:

$$\begin{aligned} S(x, y, s, x, r, b, d) &= b \text{ if } cx + sy > r \\ &= b + d \quad \text{if } cx + sy < r \\ &= 0 \text{ otherwise} \end{aligned} \tag{7-29}$$

(b) S will be determined by the method of least squares between g(x, y) and S, on surface D:

$$\int_D [g(x, y) - S(x, y, s, c, r, b, d)]^2 \, dxdy = \min \tag{7-30}$$

(c) For mathematical reasons S and F are developed in a Fourier series. The transforms are written:

$$f_i = \int_D g(x, y) \, H_i(x, y) \, dxdy \tag{7-31}$$

$$s_i\,(s, c, r, b, d) = \int_D S(x, y, s, c, r, b, d) \, H_i(x, y) \, dxdy \tag{7-32}$$

(d) Confining treatment to the first eight stages of development and finding λ_{min}:

$$\lambda(s, c, r, b, d) = \sum_{i=0}^{7} (f_i - s_i\,(c, r, s, b, d)^2) = \min \tag{7-33}$$

Hueckel's operator solves this functional equation at the expense of calculation time.

It is likely that some of the above methods could be adapted to the problems of the robot, but they are of little practical application at the moment.

7.3.2 SEGMENTATION OR CONNECTEDNESS

In detecting and storing outlines a double operation has been carried out:

(a) A new coding of the image is obtained.
(b) The image is divided into a number of lines (the outlines) and

surfaces demarcated by these lines ie the image has been *segmented.*

Although such coding and segmentation is useful it does not provide the characteristic feature required for decision-taking.

In applications where a fixed robot interacts with objects, the segmentation stage will consist of distinguishing different objects in the image. This operation is often simple if the outlines are clear and if the objects have non-overlapping images. It becomes more complicated if the objects overlap. Then resegmentation which distinguishes and separates even incompletely seen objects[92, 105] is required. A need for pattern recognition is clearly defined. For the time being suppose that this step is possible, and examine how one proceeds in the wake of good resegmentation.

7.3.3 EXTRACTION OF CHARACTERISTIC PARAMETERS

The final coding of the image, or its representation, is formed from a number of parameters evaluated for each of the objects defined by segmentation. These parameters might be the coordinates of the group of points comprising the outline, functions of the group of points representing the object or the number of intersections with arbitrary straight lines. A range of characteristic parameters that depends on the particular problem set for the robot can be invented. At present it is not known whether there exists a finite group of characteristic parameters capable of forming a universal basis applicable to all cases. Yet it seems that general parameters can be obtained that deal with present robots and robot applications, and it seems that research workers have worked in this direction, developing such parameters heuristically. There are four main classes of parameters used:

- Parameters based on geometrical considerations.
- Parameters based on functions of points in the image.
- Parameters based on topology.
- Parameters of the logical type.

7.3.3.1 Representation by geometrical features

Coordinates of the points of the image: The first geometrical representation to be considered concerns the sequence of the coordinates of points forming the surface. A variant that reduces the memory need only consist of recording the coordinates of the points of the outline. Another practical variant, for binary images, uses scanning, which must be done line by line, to record for each line the coordinates of a point and the length of segment representing the object (Figure 7.6). If a line can be read simultaneously, acquisition speed is multiplied by n. This

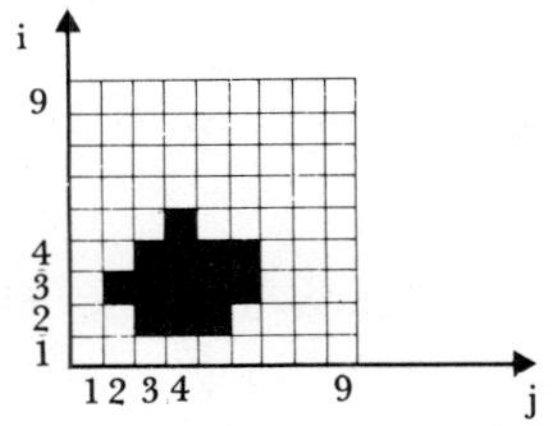

Figure 7.6. *Three modes of representation of a binary object. I_1 : 26 stored digits. I_2 : 18 stored digits. I_3 : 12 stored digits. Surface representation I_1 = (2, 3) (2, 4) (2, 5) (3, 2) (3, 3) (3, 4) (3, 5) (3, 6) (4, 3) (4, 4) (4, 5) (4, 6) (5, 4). Contour representation I_2 = (2, 3) (2, 4) (2, 5) (3, 2) (3, 6) (4, 3) (4, 5) (4, 6) (5, 4). Live length representation I_3 = (2, 3; 3) (3, 2; 5) (4, 3; 4) (5, 4; 1)*

is often possible using CCD matrices. This representation can be sufficient if there is a subsequent need to recognize the object using the masking technique (template matching).

Positional parameters: The position of the object in space leads to similar considerations to those already seen for an image on a sensitive surface. One can therefore choose a characteristic point to represent the position. This might be:

- The centre of the area of the solid image.
- An averaging point of the outline.
- The centre of the area of the minimal rectangle circumscribing the image.

Dealing with a binary image, the first two characteristic parameters can be calculated while the image is acquired. Precision is entirely dependent on the quality of detection of the outlines, but it must also be borne in mind that:

(a) The robot can operate using low precision.
(b) Following the first sighting a second, finer measurement (as the camera draws nearer to or 'zooms in' on the object) can be made, or another sensor, perhaps tactile, can complement the precise visual information required.

Orientation features of the image: The orientation of axes connected with the objects can be noted, with respect to that of the frame of the sensor. For example, the orientation of:

- The direction of the diagonals of the circumscribed, minimal rectangle.
- The direction of the 'inertial' axes of the image.

- The directions of the minimal and maximal radius vectors joining the centre of the area of the image with the outline.
- The direction of the axis joining the centre of the area of the circumscribed, minimal rectangle with the centre of gravity of the image, might be marked.

Object recognition features: Knowing the camera magnification, a number of characteristics of the image can be used for recognition purposes. Amongst these are:

- The image area (ie the number of points it contains).
- The perimeter.
- The radii originating at the centre of gravity and terminating at points on the outline; in simpler terms, the maximum radius, the minimum radius and the average radius.
- The number and the values of the maximum and minimum radii (this assumes a regular outline).
- The local extremities with respect to the grid (Figure 7.7).

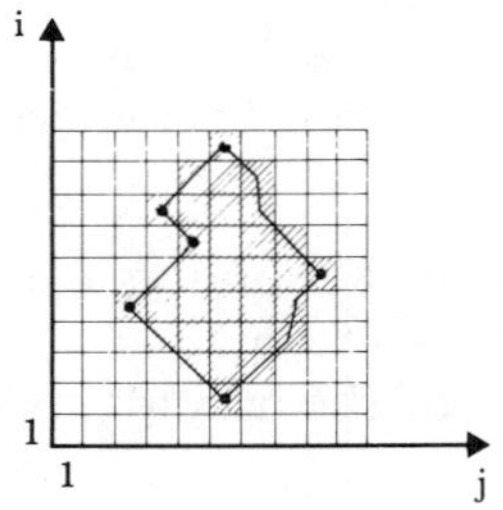

Figure 7.7. *Detection of local turn-around, (i, j) = (2, 6) (5, 3) (7, 5) (8, 4) (10, 6) (6, 9)*

- The dimensions of the circumscribed, minimal rectangle.
- The distance between the central area of the image and that of circumscribed rectangle.
- The form factor which is the ratio of the square of the perimeter to the area of the image.
- If the object contains particular characteristics, such as holes, their position and area are very significant.

If the outline is considered as a curved line the image can be characterized by:

- The curvature of each point of the outline.
- The gradient of the tangent in each point of the outline.
- The density factor, where

$$C = \int_0^S p(s)ds \tag{7-34}$$

p(s) being the curve radius at the curvilinear abscissa point s, S being the length of the outline.

If the outline is irregular, unacceptable levels of noise could result. This irregularity effect may be countered using the method described in section 7.2.2.1. Another method of regularization is obtained using hysteresis. It comprises non-linear procedures to find turn-around values of a function with real values.[86]

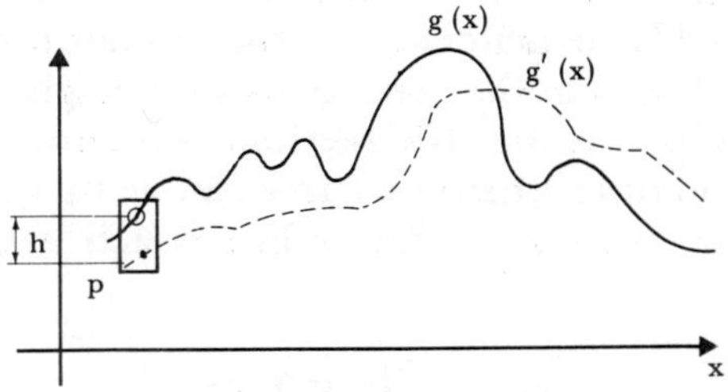

Figure 7.8. *Suppression of local turn-arounds using hysteresis*

Consider Figure 7.8 in which g(x) has to be regularized. Moving a cursor along g(x), point P only moves downwards if the cursor moves downwards. Curve g'(x) is obtained, and its shape depends only on length h, chosen at the start. The curve is thus smoothed out.

The geometrical characteristics that might be used are great in number when applied to a particular object. However, the most pertinent ones should be sought in relation to the particular problem in hand. They are likely to be less general than those listed above.

7.3.3.2 Representation by functional characteristics

As precise a description of an image as possible is needed, and this has to be created analytically, in order to provide precise and easily used information.

Fourier transform and series development: The Fourier transform (section 7.2.2.2) of image g(x, y) has already been mentioned. This is only useful in practice for outline extraction or in order to obtain correlation when instantaneous, optical means are used.[70]

However, the curve of the outline of image K(s) is in the form of a periodic function:

$$K(s) = K(s + L) \tag{7-35}$$

if L is the length of the outline. So it can be developed as a Fourier series:[106]

$$K(s) = \sum_{n=-\alpha}^{+\alpha} C_n \cdot \exp(2\pi is/L) \qquad (7\text{-}36)$$

with:

$$C_n = 1/L \int_0^L K(s) \cdot \exp(-2\pi ins/L) ds \qquad (7\text{-}37)$$

Therefore an image may be characterized (but not always an object which is three-dimensional) by the sequence of coefficients C_n.

The result is highly dependent on the curvatures encountered. If there are sharp angles, as in the vertices of a rectangle, the curve cannot be defined at these points and the method collapses. To get round this difficulty, a function other than the curve can be used, for example, the angle formed by the tangent at the outline which is the integral of the curve.

$$\theta(s) = \int_0^s K(\zeta) d\zeta \qquad (7\text{-}38)$$

The function $\theta(s) - 2\pi s/L$ is periodic but its average value is affected by the choice of reference point. The function developed in the Fourier series will therefore be:

$$\theta'(s) = \theta(s) - \mu - 2\pi s/L \qquad (7\text{-}39)$$

with:

$$\mu = 1/L \int_0^L [\theta(s) - 2\pi s/L]\, ds \qquad (7\text{-}40)$$

Given a dozen harmonic functions, any outline can be accounted for accurately. In a Fourier series, every other periodic magnitude in the image can be equally represented, particularly the function of the distance from the centre of gravity to the various points on the outline.

Polar moments:[107] Consider an image whose outline is established with reference to a set of fixed points. Let g(x, y) be the intensity of the group of points on the object. Suppose this intensity is zero away from the object. The polar moments of the image are defined by:

$$C_{nm} = \iint r^{n+1} \cos m\theta \cdot g(r\cos\theta + x_0, r\sin\theta + y_0) dr d\theta \qquad (7\text{-}41)$$

$$S_{nm} = \iint r^{n+1} \sin m\theta \cdot g(r\cos\theta + x_0, r\sin\theta + y_0) dr d\theta \qquad (7\text{-}42)$$

Where r and θ are the radius and the angle in polar coordinates with x_0, y_0 as origin.

If x_0 and y_0 are well chosen, the transformation will be invariant with respect to a translation of the image, which is important for this application. The centre of gravity of the image is often defined as the origin by:

$$x_0 = \iint x\, g(x, y)dxdy \Big/ \iint g(x, y)dxdy \qquad (7\text{-}43)$$

$$y_0 = \iint y\, g(x, y)dxdy \Big/ \iint g(x, y)dxdy \qquad (7\text{-}44)$$

It is interesting to note that the polar moments are related to the rectangular moments μ_{pq}, established in the following paragraph by relationships:[108]

$$\begin{aligned}
C_{00} &= \mu_{00} & S_{00} &= 0 \\
C_{11} &= \mu_{10} & S_{11} &= \mu_{01} \\
C_{20} &= \mu_{20} + \mu_{02} & S_{20} &= 0 \\
C_{31} &= \mu_{30} + \mu_{12} & S_{31} &= \mu_{03} + \mu_{12} \\
C_{22} &= \mu_{20} - \mu_{02} & S_{22} &= 2\mu_{11} \\
C_{33} &= \mu_{30} - 3\mu_{12} & S_{33} &= 3\mu_{21} - \mu_{03}
\end{aligned} \qquad (7\text{-}45)$$

If the image is binary and only the outline is of interest, the polar moments are simplified considerably, when:

$$C_{nm} = \sum_k r_k^{n+1} \cos m\theta_k \qquad (7\text{-}46)$$

$$S_{nm} = \sum_k r_k^{n+1} \sin m\theta_k \qquad (7\text{-}47)$$

k representing the points of the outline.

It is the same for the position of the centre of gravity:

$$x_0 = \Sigma\, i/N \qquad (7\text{-}48)$$

$$y_0 = \Sigma\, j/N \qquad (7\text{-}49)$$

N being the number of points of the outline, i the abscissa of these points and j their ordinate. These simplifications show that moments can be used for calculations in real time. The image is represented by the sequence of C_{nm} and S_{nm}.

Rectangular moments: The moment of order pq is defined by:

$$m_{pq} = \iint_{-\alpha}^{+\alpha} x^p y^q g(x, y)dxdy \qquad (7\text{-}50)$$

As for the polar moment, the series in m_{pq} determines the image g(x, y) in a unique way. It is therefore a complete description of the image that can be approached using the first few terms. Besides, it is a development in series of a generating function:

$$M(u, v) = \iint_{-\alpha}^{+\alpha} \exp(ux + vy)\, g(x, y)dxdy \qquad (7\text{-}51)$$

Taking, as before, the centre of gravity of the image as the origin:

$$x_0 = m_{10}/m_{00} \qquad y_0 = m_{01}/m_{00} \qquad (7\text{-}52)$$

One can therefore speak of *central moments* of the image:

$$\mu_{pq} = \iint_{-\alpha}^{+\alpha} (x - x_0)^p (y - y_0)^q\, g(x, y)dxdy \qquad (7\text{-}53)$$

These central moments are of some interest in robotics, thanks to the classic work of Dudani.[109] He has demonstrated that some combinations of these moments are invariant in a rotation, a translation and a similarity of the image. Here is a function which can provide recognition and the first seven combinations are as follows:[110]

$$M_1 = [\mu_{20} + \mu_{02}]^{1/2}\, B = rB \qquad (7\text{-}54)$$

$$M_2 = 1/r^4\, [(\mu_{20} - \mu_{02})^2 + 4\mu_{11}^2] \qquad (7\text{-}55)$$

$$M_3 = 1/r^5\, [(\mu_{30} - 3\mu_{12})^2 + (3\mu_{21} - \mu_{03})^2] \qquad (7\text{-}56)$$

$$M_4 = 1/r^6\, [(\mu_{30} + \mu_{12})^2 + (\mu_{21} + \mu_{03})^2] \qquad (7\text{-}57)$$

$$M_5 = 1/r^{12}\, [(\mu_{30} - 3\mu_{12})(\mu_{30} + \mu_{12})\{(\mu_{30} + \mu_{12})^2 + 3(\mu_{21} + \mu_{03})^2\} + (3\mu_{21} - \mu_{03})(\mu_{21} + \mu_{03})\{3(\mu_{30} + \mu_{12})^2 - (\mu_{21} + \mu_{03})^2\}] \qquad (7\text{-}58)$$

$$M_6 = 1/r^8\, [(\mu_{20} - \mu_{02})\{(\mu_{30} + \mu_{12})^2 - (\mu_{21} + \mu_{03})^2\} + 4\mu_{11}(\mu_{30} + \mu_{12})(\mu_{12} + \mu_{03})] \qquad (7\text{-}59)$$

$$M_7 = 1/r^{12}\, [(3\mu_{21} - \mu_{03})(\mu_{30} + \mu_{12})\{(\mu_{30} + \mu_{12})^2 - 3(\mu_{21} + \mu_{03})^2\} - (\mu_{30} - 3\mu_{12})(\mu_{21} + \mu_{03})\{3(\mu_{30} + \mu_{12})^2 - (\mu_{21} + \mu_{03})^2\}] \qquad (7\text{-}60)$$

B is an estimate of the distance between the object and the image sensor. Dealing with a binary image, calculation of the central moments is simplified to:

$$\mu_{pq} = 1/N \sum_{i=1}^{N} (x_i - x_0)^p (y_i - y_0)^q \qquad (7\text{-}61)$$

N being the number of points of coordinate image, x_i, y_i — x_0, y_0 are the coordinates of the centre of gravity. The sequence of M_i characterizes the image and lends itself very well to exploitation with the adaptive robot.[111]

It must nevertheless be noted that independence with respect to rotation, translation and similarity of relation k of the image is subject to an error owing to the discretization of the image. d is the side of a supposedly square link of the sensitive surface. These are the following errors:[112]

$$\frac{\Delta M_1}{M_1} \leqslant \frac{B^2}{2M^2}(\mu_{10}+\mu_{01})(K+1)d \tag{7-62}$$

$$\frac{\Delta M_2}{M_2} \leqslant \frac{4\Delta M_1}{M_1} + \frac{4B^4}{M_1^4 M_2}[(\mu_{02}-\mu_{20})(\mu_{01}-\mu_{10}) + 2\mu_{11}(\mu_{01}+\mu_{10})](K+1)d \tag{7-64}$$

$$\frac{\Delta M_3}{M_3} \leqslant \frac{6\Delta M_1}{M_1} + \frac{(K+1)dB^6}{M_1^6 M_4}[(\mu_{30}-3\mu_{12})(3\mu_{20}-3\mu_{02}-2\mu_{11}) + (\mu_{21}+\mu_{03})(3\mu_0+\mu_{02}-2\mu_{11})] \tag{7-64}$$

$$\frac{\Delta M_4}{M_4} \leqslant \frac{6\Delta M_1}{M_1} + \frac{(K+1)dB^6}{M_1^6 M_4}[(\mu_{30}+\mu_{12})(3\mu_{20}+\mu_{02}+2\mu_{11}) + (\mu_{21}+\mu_{03})(3\mu_{02}+\mu_{20}+2\mu_{11})] \tag{7-65}$$

$$\frac{\Delta M_5}{M_5} \leqslant \frac{12\Delta M_1}{M_1} + \frac{(K+1)dB^{12}}{M_1^{12} M_5} f_5 \tag{7-66}$$

$$\frac{\Delta M_6}{M_6} \leqslant \frac{8\Delta M_1}{M_1} + \frac{(K+1)dB^8}{M_1^8 M_6} f_6 \tag{7-67}$$

$$\frac{\Delta M_7}{M_7} \leqslant \frac{12\Delta M_1}{M_1} + \frac{(K+1)dB^{12}}{M^{12} M_7} f_7 \tag{7-68}$$

with:

$$f_5 = \Delta[(\mu_{30}-3\mu_{12})(\mu_{30}+\mu_{12})\{(\mu_{30}+\mu_{12})^2+3(\mu_{21}+\mu_{03})^2\} + (3\mu_{21}-\mu_{03})(\mu_{21}+\mu_{03})\{3(\mu_{30}+\mu_{12})^2-(\mu_{21}+\mu_{03})^2\}] \tag{7-69}$$

$$f_6 = \Delta[(\mu_{20}-\mu_{02})\{(\mu_{30}+\mu_{12})^2-(\mu_{21}+\mu_{03})^2\} + 4\mu_{11}(\mu_{30}+\mu_{12})(\mu_{21}+\mu_{03})] \tag{7-70}$$

$$f_7 = \Delta[(3\mu_{21}-\mu_{03})(\mu_{30}+\mu_{12})\{(\mu_{30}+\mu_{12})^2-3(\mu_{21}+\mu_{03})^2\} - (\mu_{30}-3\mu_{12})(\mu_{21}+\mu_{03})\{3(\mu_{30}+\mu_{12})^2-(\mu_{21}+\mu_{03})^2\} \tag{7-71}$$

When researching functional representation of the image, many other developments (eg Tchebychef polynomials) may be used. It is important that the development be mathematically valid, that is can characterize the image in a unique way (is possible independently of the position of the image on the sensor) and that, finally, it can be evaluated in real time using a reasonable amount of computation.

7.3.3.3 Representation by topological features

Topological features are those that represent regions in space. The aim is to encode their positional relationship and other similar properties:

(A) The resolution of the image into a series of connected regions is not necessarily easy. This is demonstrated in Figure 7.9. Assuming that the eight neighbouring points of a given point are connected to it, the 1s appear connected together but the region inside that defined by 1s is connected with the exterior region. If a four connection is used, the region inside the 1s is no longer connected with the exterior region, but the 1s form four distinct regions.

0	0	0	1	1	0	0	0	0
0	0	1	0	0	1	0	0	0
0	0	1	0	0	1	0	0	0
0	0	0	1	1	0	0	0	0
0	0	0	0	0	0	0	0	0

Figure 7.9. *The connectedness problem*

Mathematically, region R is called a *group of points* with the following properties:

(a) All points in R have the same grey level.
(b) Any two points in R can be connected by a chain of adjacent points all situated in R.
(c) Not every group of points containing R satisfies properties (a) and (b).

Using this definition an image can be resolved into regions that almost follow the outlines of the objects.

(B) Labelling each region can be done by assigning any name simply so as not to confuse the detected regions; assigning a name based on a quality of the region or assigning a name that corresponds with the geometrical figure by which the region can be identified. In this last case the recognition process will not often be encroached upon.

(C) Coding topological information can be of several levels: between the regions or enumeration of types of regions. For example, in a flat figure, one can represent it topologically by its departures from convex. Recalling that a figure is called convex if each segment of the straight line joining two of its points is contained in this figure. Consider an arbitrary figure S. The smallest convex group containing X (Figure 7.10) is called *Hull's convexity*. One can characterize S by its convex deficiencies D_i, comprising two 'bays' and two 'lakes', which is a simpler description than the description of S itself.

When topological coding takes into account positional relationships of regions, linguistic methods are often used.

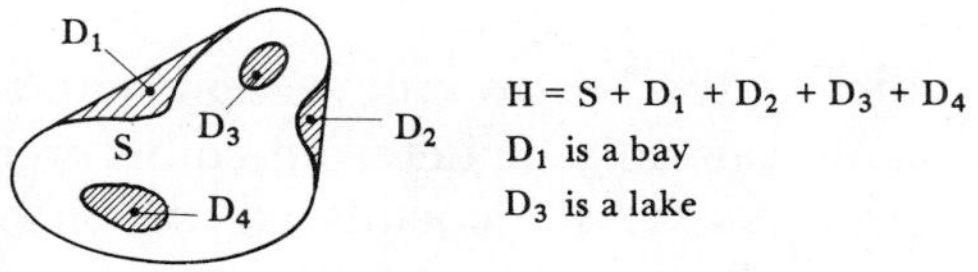

Figure 7.10. *Hull's convexity*

7.3.3.4 Skeletonization[113]

This procedure transforms an image with a particular surface into an image formed by a few lines that represent its skeleton. If the rules of formation of the skeleton are well chosen, information about the shape of the primitive image is always present in this skeleton. This procedure has been much used in the recognition of hand-writing.

The best known technique has the technical name *prairie fire*. Consider an image such as that shown in Figure 7.11. Suppose the interior is made of dry grass and that the exterior is composed of wet grass that does not burn. Setting fire simultaneously to the outline of the whole figure, it is supposed that it spreads at constant speed towards the centre of the figure. The points where the fire is extinguished through the encounter of two burning fronts, form the skeleton of the figure. For example, for a circle the skeleton would be formed at one point. Mathematically, the skeleton of a figure is defined in the following way.

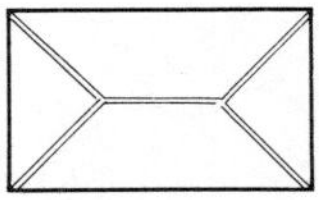

Figure 7.11. *Skeletonization of a rectangle*

The distance of a point x to a group of points A forming the outline of an image is the size:

$$d(x, A) = \text{minimum}[d(x, y), y \in A] \qquad (7\text{-}72)$$

Some points situated in the interior of the figure are such that y is not unique. For these points:

$$d(x, A) = d(x, y) = d(x, z) \qquad (7\text{-}73)$$

These points x define the skeleton of A. There are reconstruction procedures for A starting with the skeleton, which show that the information in form A is preserved.

Skeletonization is therefore a transformation of an image comprised of points which it will have to characterize, in its turn, by topological, geometrical or functional considerations.

7.4 Choice of representation

The question is, which is the best of these representation procedures? For the robot this cannot usually be determined beforehand because, as has already been emphasized, it depends on the information that is available and the purpose for which it is required.

Generally, an image will not be represented by a single representation, but by a number of complementary representations. In some cases the surface and the polar moments would be used, for others form factor, the centre of gravity of the image and the central moments.

Attention should be drawn to the relationship between the form of representation chosen and strategy chosen to acquire the pertinent information.

The choice of representation will be made as a function of the observation strategy and vice versa. In the design stage, one can choose either a constraint on the strategy (eg choose a fixed 2D camera) and choose a representation that is an adequate match, or choose constraints in representation and choose an acquisition strategy providing the information needed.

Flexibility of the sensor would be the greatest asset, and representation is the simplest. Yet a very important criterion is that of the execution time and the choice must be optimized first, using the criterion: will a single frame using complex calculations take longer than two or three frames using simple calculations?

Conclusions

In this chapter the problem of representation of information contained in an analog image produced by a sensor in a form capable of being handled by a computer has been briefly examined. How to condense the information has been considered, eliminating, as much as possible, imprecisions and obtaining general methods applicable to all classes of objects. Representation methods can be quite varied – the treatments can be analog, optical or digital. Whilst it cannot be definitely stated which is the best representation system for robotic vision, a combination of available methods, depending on the particular application, is at our disposal.

Chapter 8

Remote sensing: artificial vision—exploitation of information contained in the representation of images

In previous chapters an attempt has been made to discover how a robot may obtain and use images to form a multi-dimensional data base which will facilitate the extraction of pertinent information. The first three stages shown in Figure 8.1 have now been covered. Now, the robot has to act on information contained in the representation stage so as to

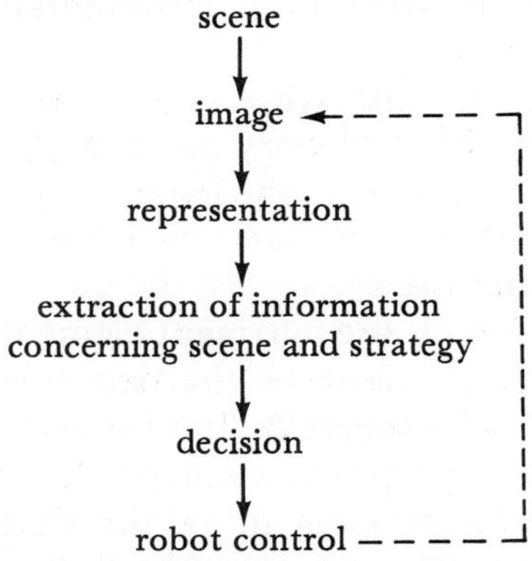

Figure 8.1. *Major steps in visual interaction*

take decisions about its behaviour and to take the process further so as to reach the point where the robot can perform:

1. Unlike the constraints presented by 'classical' pattern recognition, the robot can increase the richness of its information using successive frames. In other words, it is possible to have two kinds of information in one frame: one provides information about how to take the next frame. This is an essential aspect of environment-robot interaction. The message obtained from the scene is not only *descriptive* but it is also an *active*, interactive or strategic message.

In examining the different possible representations discussed in Chapter 7 (which all result from the use of 'classical' methods), it becomes obvious that the descriptive message initiates the process. The active message can only be deduced by using an algorithm operating on the descriptive representation. This accords with the theory of human vision proposed by Helmholtz and discussed in Chapter 6.

2. The robot does not have to rely upon visual sense alone in carry-

ing out a given task (although when using mobile cameras equipped with zoom, this situation is approached).

3. Just as the representation of an image assumes a prior decision regarding its coding, the process of recognition also assumes that knowledge has already been obtained in the learning phase.

8.1 Task representation and learning

In designing a robot with vision, the nature of the task will determine the choice of pertinent elements in the visual presentation that the robot will be called to act upon. The tasks, their mode of execution and the pertinent elements to be used by the robot will be laid down by the designer. He then formulates representations of the scene derived from the images available, allowing for easy access to frame features. The features do not add up to a total representation of the scene, but simply are contained in it.

This is even more the case when a more general representation is needed ie when the robot has to carry out a range of tasks. For example, suppose the pertinent element is the length of the diagonal of a square. When other tasks or imperatives are taken into account, this might be changed to representing the square by the coordinates of two extremities of the square. To derive the pertinent element ie the diagonal of the square, again, the distance between the two points represented by the coordinates will have to be found and multiplied by $\sqrt{2}$.

Having obtained the pertinent element the decision stages will be reached, but the pertinent element ie the diagonal rather than the characteristic magnitude (the side length) will first be needed.

Having chosen an appropriate representation, the operator has to teach it to the robot, but it has to be spelled out for the scenes that the robot will have to examine (Figure 8.2).

Referring to Figure 8.2, the learning phase could be envisaged either as an evaluation of pertinent elements or the implementation of decisions that follow it. But in practice that would create complexities in the calculation stages.

In present practice, the learning phase consists of feeding reference values of the elements that form the data base of the representation chosen for the image into the computer.

However, one may be interested in a number of objects simultaneously. So one feeds in data about each of the objects, knowing that, if the total scene is comprised of a number of objects that need to be retrieved, data on all these possible combinations must also be supplied.

Several efforts are being made to tackle this problem.[92, 105, 108] However, much research is still needed.

Once a mode of representation has been chosen, two procedures can

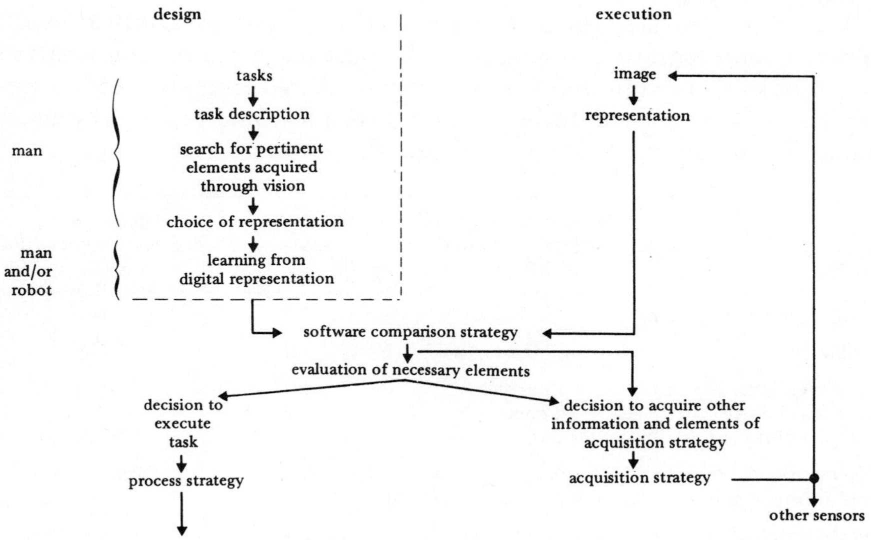

Figure 8.2. *The principle of use of vision*

be used in the learning phase:

1. (a) If a quasi perfect, mathematical model of the object is available, and the magnification due to the optical sensor is known, data can be fed into this model, and this will generate the representation algorithm and will allow calculation of the value of the selected features.
   (b) If characteristic parameters are directly available for use as pertinent data (number of peaks, number of holes etc) they can be fed in directly.
2. If the range of objects is unrestricted, which is always the aim, the mathematical model not only becomes complex but also inaccurate. A satisfactory solution would be to photograph the objects in question with the optic sensor itself involved, and retain the results in memory. They will then form the reference representation or learned representation.

8.2 Reference representation

The amount of reference representation required is completely different for flat two-dimensional objects or three-dimensional objects. It also depends on the information provided beforehand, on the characteristics of the sensor and on whether the 2D, K-2D or 3D technique is used. It can be characterized by the number of images that need to be captured to create the reference representation.

Here too, representation of an *image* must not be confused with reference representation of an *object.* To have access to the information on an object, a large number of images taken from different angles (section 6.5) may need to be formed in the learning phase. The most common cases are summarized in Figure 8.3:

	number of images per object to create the reference	visual technique	number of images needed in recognition stage
two-dimensional objects	1 image	2D	1
flat object	1 or 2 (front/back)	2D	1
three-dimensional object placed on a known horizontal plane	1 per stable support side	2D	1
three-dimensional object placed in any way	theoretically infinite (in practice a few hundred)	K-2D	from 1 to η
three-dimensional object placed in any way	theoretically infinite (in practice a few)	3D	1

Figure 8.3. *The number of registered images as a function of the nature of the object, its relationship with the environment and the visual technique to be used*

1. With a two-dimensional object it is sufficient to record one image, a group of characteristic vectors written as:

$$V^1_{Ref} = [(a^1_1, a^1_2 \ldots a^1_n), (b^1_1, b^1_2 \ldots b^1_m) \ldots (\ell^1_1, \ell^1_2 \ldots \ell^1_q)] \quad \text{(8-1)}$$

in which $(a^1_1, a^1_2 \ldots a^1_n)$ are the components of the first characteristic vector etc whilst $(b^1_1, b^1_2 \ldots b^1_m)$ are components of the second characteristic vector etc with their numerical reference values.

For example, if the image is represented by the surface parameters of the image (S), form factor (F), number of peaks (N), the combinations of the Dudani central moments up to the fifth (M_i) and the first four elements in the Fourier series for the outline (C_j), the characteristic group:

$$V = [(S), (F), (N), (M_1, M_2, M_3, M_4, M_5), (C_1, C_2, C_3, C_4)] \quad \text{(8-2)}$$

from the frame the reference will be obtained:

$$V^1_{Ref} = [(S^1), (F^1), (N^1), (M^1_1, M^1_2, M^1_3, M^1_4, M^1_5), (C^1_1, C^1_2, C^1_3, C^1_4)] \quad \text{(8-3)}$$

2. In the case of a flat object, a photograph for each side, if one is not a mirror image of the other, can be recorded.

3. For three-dimensional objects an often-used technique is to place the object on a horizontal plane and take a limited number of frames. By taking a photograph along an axis perpendicular to the support plane, the object is represented by as many two-dimensional objects as it possesses support sides. This procedure solves a large category of problems, but is not of universal application. This can easily be demonstrated.

Consider the pyramid in Figure 8.4. If a photograph is taken along

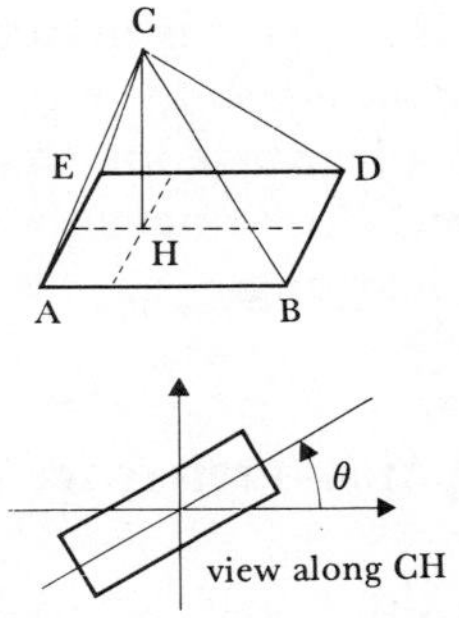

Figure 8.4. *θ cannot be computed with accuracy greater than π with view from above*

direction CH, a rectangle is obtained whatever the position of C, providing its projection H is within ABDE. In these conditions, the orientation of AB or ED up to values of 180° only can be found. Such a technique will not only provide information concerning bulk where the object is supported by the horizontal plane only, but in any plane.

4. To obtain information independent of the nature of the object and so as not to be restricted by its position in a horizontal plane, several two-dimensional images can be taken. This is the K-2D technique. As nothing is known about relative orientations of object and camera, this assumes that the characteristic reference vector for all possible object-camera orientations has been recorded. When the camera is fixed and the object rotated in front of it, there are three possible rotations (Figure 8.5) each of 360°. The number of photographs to be taken

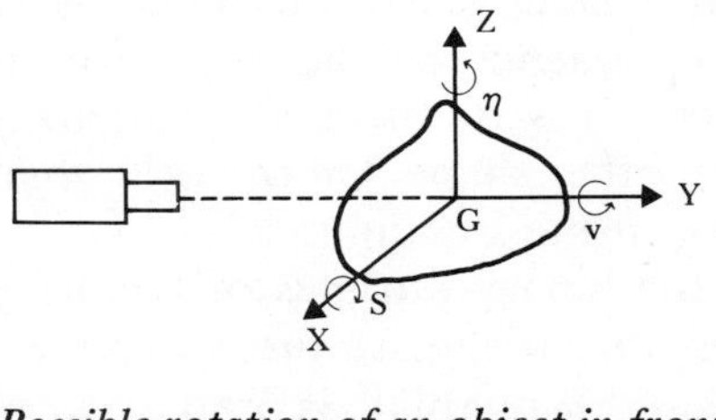

Figure 8.5. *Possible rotation of an object in front of a camera*

depends, therefore, on the rotations chosen and the qualities of the object (symmetry: one photograph is sufficient for a sphere, the pyramid in Figure 8.4 requires images up to a symmetry plane, rotations of 180°).

So far as the characteristic parameters are concerned, taking Dudani moments for example, they are invariant in a rotation of the image on itself. From Figure 8.5 it is seen that the shots corresponding to rotation v are useless. In the worst case, there are $(360/p)^3$ reference vectors to be registered, p being the *sampling step in degrees*, assumed to be the same for the three possible rotations.

5. Using a 3D sensor the flat image is matched by a knowledge of the position of a point on the image and a corresponding point on the object. This is clearly more advantageous than using two flat images. If the object does not demand too fine an examination, two or three photographs are enough for learning.

8.3 Comparison between input and reference representation

Every comparison assumes the existence of two entities and a way of evaluating their difference. Here, the difference will be termed the *distance* between these two entities. Representation can take two different main forms:

1. A group of points forming the image, represented by their coordinates (or a transformation of the image, such as its skeleton).
2. A group of characteristic values of this group of points, logical, topological, geometrical or functional magnitudes.

Although one can see clearly what constitutes the second group, nothing has been said about representations that can be used for an image formed of points. There are two main types of such representation: masks (templates) or models of the image formed by intersections with arbitrary curves.

8.3.1 TEMPLATE MATCHING

Template matching consists of making two images of sets of points coincide. If coincidence takes place, the object is recognized. This has the advantage of independence of the object or its complexity. In a two-dimensional system, it is no more difficult to recognize a key outline than a disk or any other shape. On the other hand, this method can be very time-consuming using a computer.

Figure 8.6 shows the binary outlines of a cube and pyramid. Using coincidence of images, one wishes to discover whether this figure contains a square. So a mask or template is prepared, composed of an array

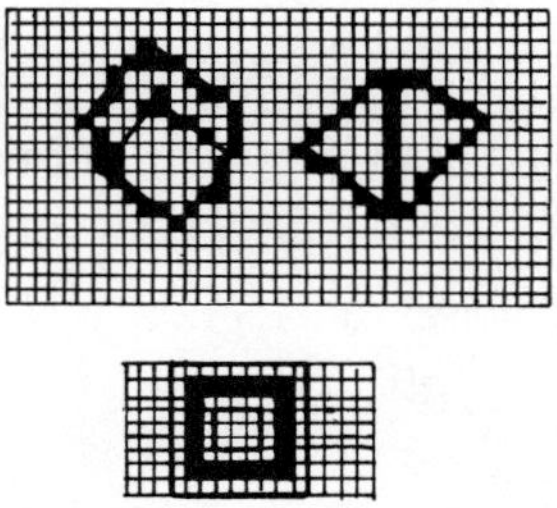

Figure 8.6. *Template matching*

of *black points* forming a square, surrounded by an enclosing array of *white points*. This mask is to be applied to the figure and moved until coincidence occurs. There are several difficulties:

1. To use the mask, it must be of the right size. If it is not, a whole series of masks will be needed.
2. In Figure 8.6, the hypothetical square can be situated anywhere and with any orientation. So double scanning is needed: for translation and rotation.
3. Figure 8.6 demonstrates the effect of spatial quantization. Even if the mask is placed correctly, coincidence will not be perfect but merely the best possible.

This is why the mask technique will only be useful if information is available which allows one to know beforehand the required dimensions of the mask, thus reducing the number of possible comparisons to a reasonable number.

How is the distance between the image and the mask estimated?

Let g(i, j) be the image to be examined; t(i, j) the mask and D the field of the mask's definition. The distance can be expressed by:

$$M(m, n) = \sum_i \sum_j |g(i, j) - t(i - m, j - n)| \tag{8-4}$$

or, in Euclidian measure:

$$E(m, n) = \left[\sum_i \sum_j \{g(i, j) - t(i - m, j - n)\}^2\right]^{1/2} \tag{8-5}$$

or, with measure of norm $< \infty$:

$$S(m, n) = \max_{i,j} |g(i, j) - t(i - m, j - n)| \tag{8-6}$$

In these three expressions, i and j vary as (i − m) and (j − n) remain within D. For coincidence the functions M, E, S each takes up a minimum value.

Examining the Euclidian norm a little closer:

$$E^2(m, N) = \sum_i \sum_j [g^2(i, j) - 2g(i, j) \cdot t(i-m, j-n) + t^2(i-m, j-n)] \tag{8-7}$$

When i and j are varied, whatever m and n, t^2 corresponds with the number of points of the mask and remains constant.

On the other hand, if the mask is placed on the image at random, the sum of the intensities of the points covered by the mask is constant:

$$\sum_i \sum_j g(i-j) = \text{constant} \tag{8-8}$$

with i and j belonging to points situated under the mask.

In equation (8-7) there is a variable monomial which is none other than the function of correlation between g and t:

$$C_{gt} = \sum_i \sum_j g(i, j) \cdot t(i-m, j-n) \tag{8-9}$$

(like $i-m, j-n, \epsilon D$)
C_g appears with the sign less in equation (8-7). The best coincidence will take place when C_{gt} is maximum.

To remove errors due to the hypothesis of the constancy of $\sum_{i,j} g(i, j)$, equation (8-9) can be normalized:

$$R_{gt} = C_{gt} / \left[\sum_i \sum_j g(i, j)^2 \right]^{1/2} \tag{8-10}$$

i and j always verifying $i-m, j-n, \epsilon D$.

The mask technique can be useful in specific cases. It has been employed outside robotics using optical methods, because optical correlators can be made.[72,114] However, in these cases the dimensions of the mask to be used are always known, and the objects are correctly presented in orientation and position. The only objective is recognition of an object among a group. The robot has to do more.

8.3.2 COMPARISON WITH INTERSECTION MODELS

In the intersection method deterministic or statistical models of relationships between the images and figures that are derived from them are used. These figures might be masks, but are preferably simpler. Generally, straight lines or circles are used. The intersections of these straight lines (or circles) are related to the outlines of the image.

The straight lines are drawn on the image surface (Figure 8.7). The number of straight lines that do not intersect with the figure, as well as the distribution of the lengths of the segments cut by the straight lines in the figure, can characterize the image. Yet, the image constructed by the use of intersections must be invariant in rotation and translation of the optimal image.[115] Using straight lines of fixed direction, indication of the object's orientation could be obtained.

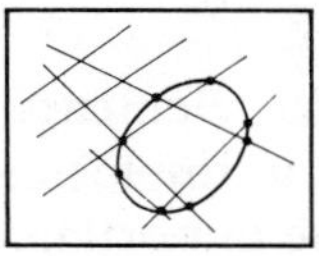

Figure 8.7. *Recognition by means of arbitrary lines*

This method is over-sophisticated for use in the type of robot that interests us. The same information can be obtained using simpler means, based on stored information.

8.3.3 DISTANCE BETWEEN INPUT AND REFERENCE

Referring to section 8.2, it will be recalled that the representation of the image can be provided by a group of characteristic vectors:

$$V^k_{Ref} = (a^k_1, a^k_2, \ldots a^k_n), (b^k_1, b^k_2, \ldots b^k_m) \ldots (\ell^k_1, \ell^k_2 \ldots \ell^k_q) \qquad \textbf{(8-11)}$$

where k = 1 to t. A frame provides an image with which the group of characteristic vectors can be calculated:

$$V = (a_1, a_2, \ldots a_n), (b_1, b_2, \ldots b_m) \ldots (\ell_1, \ell_2, \ldots \ell_q) \qquad \textbf{(8-12)}$$

It is a matter of knowing the distances between V and V^k_{Ref} (t in number), of selecting one or more distances and of making decisions based on these minimal distances.

The first type of decision arises when action results directly from information of the deciphered image, as well as from supplementary information. The second type of decision arises when the image provides insufficient information and more has to be sought, both visually and otherwise. The third type of decision arises when the process is halted due to insufficient information. Adequate feedback (which must not be unstable) must be provided to stop the procedure from entering an endless cycle.

The estimation of the distance between measurement and reference is therefore important and has direct consequences on the effectiveness of the system. The estimate must be made considering that:

1. It is doubly multi-dimensional.
2. The matrix

$$V = \begin{pmatrix} V_a \\ V_b \\ V_c \end{pmatrix} = \begin{pmatrix} a_1, a_2, \ldots a_n \\ b_1, b_2, \ldots b_m \\ \ell_1, \ell_2, \ldots \ell_q \end{pmatrix} \qquad \textbf{(8-13)}$$

is not homogeneous, because it comprises both analog and logical elements.

3. *The reference* for the analog elements may be imprecise, but not for the logical elements.
4. *Measurement* of the analog elements may be imprecise and a probability of error may arise in the logical elements.
5. There is a hierarchy in the importance of the measurements:
   (a) A hierarchy for the vectors $V_a V_b \ldots V$.
   (b) A hierarchy for the components of these vectors.

   This hierarchy is based on two criteria:

   (a) The richness of information supplied by a particular element with respect to the task.
   (b) The importance of the decision to be taken after the measurement. A decision based on erroneous information could be disastrous, and too long a delay in verifying the information would render the decision-making useless.

So any type of distance shown in the matrices L^1, L^2, L^∞ can be chosen according to equations (8-4) to (8-6). The problem arises in the estimation of proper balance to obtain a generalized distance between measurement and reference, requiring the use of a simple discriminant, and especially an effective function (Chapter 9). This estimation is in reality often made heuristically, but any of the learning methods discussed by Duda and Hart[86] could be used to determine it.

That distance may be expressed in Euclidian measure. To faciltate this, V and V^k_{Ref} are written in the following form:

$$V = \begin{pmatrix} V^1 \\ V^2 \\ V \end{pmatrix} = \begin{pmatrix} a^1 & a^2 & \ldots & a^1_n \\ a^1_2 & a^2_2 & \ldots & a^2_{n_2} \\ a^\ell_1 & a^\ell_2 & \ldots & a^\ell_{n_1} \end{pmatrix} \tag{8-14}$$

$$V^k_{Ref} = \begin{pmatrix} a^{1k}_1 & a^{1k}_2 & \ldots & a^{1k}_{n_1} \\ a^{2k}_1 & a^{2k}_2 & \ldots & a^{2k}_{n_2} \\ a^{\ell k}_1 & a^{\ell k}_2 & \ldots & a^{\ell k}_{n_3} \end{pmatrix} \tag{8-15}$$

where k = 1 to t

There will be a proper balance per line of V, g_{jk}, and a proper balance inside an identical line g_{ijk}. The distance normalized between V and V^k_{Ref} may be written (by choosing least squares):

$$d^*_k = \frac{\sum_{j=1}^{\ell} g_{jk} \sum_{i=1}^{n_j} g_{ijk} (a_i^{j^2} - a_i^{jk^2})}{\sum_{j=1}^{\ell} g_{jk} \sum_{i=1}^{n_j} g_{ijk}} \tag{8-16}$$

If all the elements contribute in the same way to the information wanted, which is generally assumed, weighting proportional to the relative precision of the measurement of each a_i^j for g_{ijk} and each V_j for g_{jk} should be applied.

For example, using representation via Dudani moments only:

$$V = (M_1, M_2, M_3, M_4, M_5)$$

In section 7.3.3.2 an expression for error due to spatial quantization $\Delta M_i/M_i$ was used. Suppose, arbitrarily, that this is the only source of error. As V is only formed by one vector, $g_{jk} = 1$ is still obtained given that the precision for each registered V^k_{Ref} is identical. Otherwise g_{jk} is a function of k, and the g_{ijk} values will be inversely proportional to $\Delta M_i/M_i$. (The distance is normalized and the weight of the precise terms must be more important than those of the imprecise terms.) Therefore the distance is:

$$d^*_k = \frac{\sum_{i=1}^{5} (M_i/\Delta M_i)(M_i^2 - M_i^{k^2})}{\sum_{i=1}^{5} M_i/\Delta M_i} \tag{8-17}$$

In this example the weighting could be calculated on line, for each image frame, beginning with equations (7-56) to (7-65). But it should not be forgotten that these calculations may be protracted and that the distances derived serve to make decisions. So they are exploited using a logic, and it is better to make a minimal over-estimation of errors and keep the g_{ijk} values fixed, except in specific cases where adaptivity would be both easier and/or essential.

8.4 Discriminant functions

Assume that the distances between the series of reference representations of the scene and the representation obtained from image-taking have just been calculated. The series of distances between V and V^k_{Ref} is available ie:

$$d^* = (d_1^*, d_2^*, \ldots d_k^* \ldots d_t)$$

d* must be examined and a first decision taken. Can V be attributed to one value of the V^k_{Ref}? In principle, the minimal distance is that which indicates that V is the nearest to V^k_{Ref}, and it could be decided that V is identical to V^k_{Ref}. In reality several elements must be taken into account:

1. Has the weight of d_k been chosen correctly, so that the smallest distance coincides with the strongest probability that V is identical to V^k_{Ref}?
2. How does one interpret all seemingly 'very large' distances?
3. How can the fact that there are several distances very close to the minimum distance be explained? In other words, how much can one rely on $d_{k,min}$ and what risk is taken in deciding to identify V with V^k_{Ref} ($d_{k,min}$)?

8.4.1 THE BAYESIAN DECISION

This is based on the probability of a cause given a known effect. This simply means that, after measuring V, this value is going to be attributed to V^k_{Ref} if the probability of obtaining V^k_{Ref}, knowing V, is maximum. This probability is called the *a posteriori conditional probability,* written as:

$$P(V^k_{Ref} \mid V)$$

How can this probability be estimated?

Imagine an object is to be identified using the technique of moments, as in the example discussed in section 8.3. A group of photographs of the object is taken that corresponds to all the possible orientations relative to the camera. Call this number t. Taking an image of the object for recognition purposes, there is one chance in t that one might 'stumble' on one such value chosen in advance. This is the *a priori* probability of finding V^k_{Ref}. This probability is written $P(V^k_{Ref})$.

Now suppose the object is placed so that the camera views it in the direction which corresponds to V^k_{Ref} and that a large number of photographs is taken, from which V is calculated every time. Because of the various sources of error, for example slight variations in illumination or image noise, $V = V^k_{Ref}$ would not be obtained always, but a certain distribution of V centred on V^k_{Ref} (Figure 8.8) would be obtained.

For another V^k_{Ref} another distribution could be obtained. If the same number of photographs for each V^k_{Ref} is not taken, valid comparisons cannot be made. This is why this distribution is normalized. It is called the *conditional probability density*. It is written:

$$P(V \mid V^k_{Ref})$$

The probability density of V can also be written:

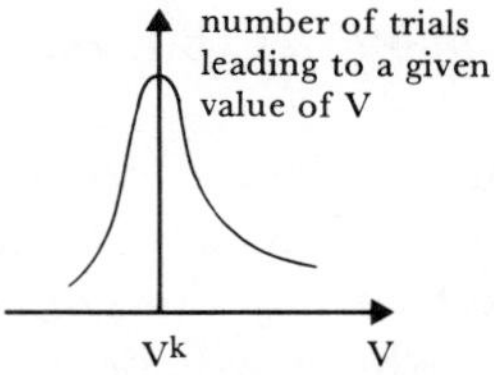

Figure 8.8. *Distribution of V for a number of trials that should yield* $V = V^k_{Ref}$

$$p(V) = \Sigma p(V \mid V^k_{Ref}) \cdot P(V^k_{Ref}) \tag{8-19}$$

This is the sum of each probability density of V taking frames that correspond to V^k_{Ref}, multiplied by the probability of having V^k_{Ref} (Figure 8.9).

The Bayesian rule provides a relationship between all these probabilities and probability densities:

$$P(V^k_{Ref} \mid V) = \frac{P(V^k_{Ref}) \cdot p(V \mid V^k_{Ref})}{p(V)} \tag{8-20}$$

After measuring V, the probability of obtaining V^k_{Ref} is equal to the *a priori* probability of having V^k_{Ref}, multiplied by the ratio of the probability density of V related to frames meant to be V^k_{Ref}, to a probability density V.

The three terms in the right-hand side of the equation (8-20) are terms that can be set in advance (at least theoretically). So for each input frame to be recognized, the probability $P(V^k_{Ref} \mid V)$ can be calculated.

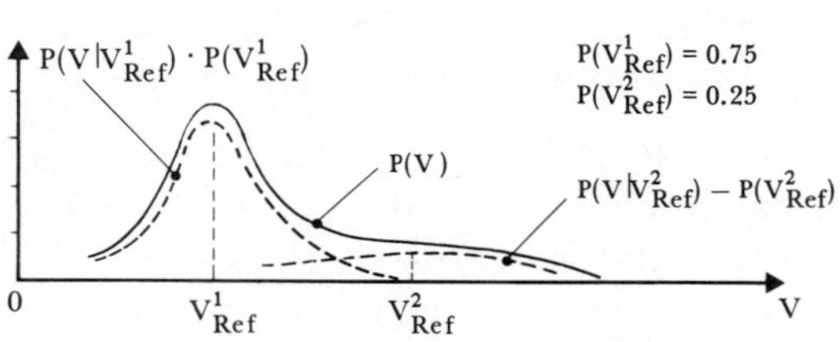

Figure 8.9. *Example of the build-up of P(V) with two a priori reference probabilities 0.75 and 0.25*

In using probabilities, one risks errors, all the more so because a number of probabilities can be very close. So the decision is not going to bear on *a posteriori probabilities*, but on the risk one takes in attributing V values to V^k_{Ref}.

The recognition procedure will be followed by a number of possible actions. For robots, these actions are divided into two categories: informative actions and executive actions. Each category will comprise several possibilities, so that there will be w different possible courses of action in total, which will be called $\alpha_i, \alpha_2 \ldots \alpha_w$.

If a wrong attribution of V to V^k_{Ref} is made and if that triggers off action α_i, a mistake will be created which could be either minor (slight extension of the task) or serious (wrong manoeuvre causing damage).

Calling $\lambda(\alpha_i \mid V^k_{Ref})$ the *loss* if action α_i is engaged having attributed V to each of reference V^k_{Ref}:

$$R(\alpha_i \mid V) = \sum_{k=1}^{t} \lambda(\alpha_i \mid V^k_{Ref}) \cdot P(V^k_{Ref} \mid V) \qquad \textbf{(8-21)}$$

This risk is called the *conditional risk*. At each frame, the risk to be taken for each possible action will be estimated, and the action which entails the smallest conditional risk will be chosen. This minimum conditional risk is called the *Bayesian risk*.

By using $R(\alpha_i \mid V)$ a *classifier* ie a 'machine' which calculates functions $g_i(V)$ and which will decide action α_i can be constructed if:

$$g_i(V) > g_j(V) \qquad \textbf{(8-22)}$$

It has been seen that action α_i is taken if $R(\alpha_i \mid V)$ is a minimum, so that the definition of the classifier requires equation (8-22); then:

$$g_i(V) = -R(\alpha_i \mid V) \qquad \textbf{(8-23)}$$

This is a *Bayesian classifier*. Function $g_i(V)$ is called the *discriminant function*.

Note that the decision does not change if, instead of taking $g_i(V)$ as discriminant function, $f[g_i(V)]$, an increasing and monotonic function of $g_i(V)$ is taken.

For robots, these probability density functions are often not known. If one wants nevertheless to use the Bayesian classifier, an estimate of it must be made, and two approaches are possible:

1. The *parametric method* in which it can be supposed that the density function can be characterized by a finite number of parameters. If Gaussian distributions are used (as is often done), these parameters are the average and the variance. This method imposes a form based on an analytic function, and this is not always possible.
2. *Non-parametric methods* which can be used without prior knowledge of the probability density. The estimates are made by

counting[116,117] using the nucleus method,[118–120] by the 'nearest neighbour' methods[121] and others ie the methods are all based on measurements in the learning phase which will have to be repeated for each robot task. As the modelling can never be perfect, the robot will rarely attain the performance required for industrial use, despite the sophisticated methods used. This is why the Bayesian classifier, handicapped by the need to estimate probability densities, remains in the experimental stage, so far as its application to industrial robots is concerned.

8.5 Distance and attribution

Returning to equation (8-18):

$$d^* = (d_1^*, d_2^*, \ldots d_k^*, \ldots d_t^*)$$

expresses the distances between the measured representation V and the reference representations V_{Ref}^k, (k = I, t).

A discriminant function and a classifier must be found. The Bayesian classifier is optimal ie it minimizes the risk. To improve its application there are two solutions:

1. Transform the probabilities into certainties.
2. Use it in a sequential way to make the attribution probabilities converge towards certainty.

The first solution depends only on technological data of known limits. Certainty can only be obtained in simple cases where environmental constraints hardly compatible with the objectives for the robot are accepted.

So, generally, the second solution with non-Bayesian classifiers, using determinist decision trees, will be chosen. But use of the decision tree can require most complex techniques (page 214). Moreover, this greatly simplifies learning the decision rule, which becomes limited to threshold adjustments.

It will be supposed that to establish equation (8-18) the weighting factors have been chosen correctly (although heuristically), so that the error made in attributing V to V_{Ref}^k corresponding to minimal number of d_k remains statistically 'weak'.

In other words, for a large number of tests involving each V_{Ref}^k, a percentage with poor recognition (20 to 30 per cent), but compatible with the results derived, will be obtained. This indicates that distance has significance and that its choice is not arbitrary.

In these conditions the success rate would be greatly improved by using discriminant functions. These are reduced to distance thresholds (or to one threshold), but other information using other sensors or camera angles might also be used.

A first method consists of choosing a threshold (which will have to be improved having seen the results) and keeping all distances below this threshold. There are three possible cases:

– No distance is less than the threshold. This will be diagnosed as poor setting of the threshold.
– A single distance is below the threshold. V is then attributed to the corresponding V^k_{Ref}.
– Several distances are below the threshold. In these conditions a second image is taken, keeping only selected representative V^k_{Ref} as reference representation. Successive frames will diminish V^k_{Ref} until there is only one left. V is attributed to this V^k_{Ref} and executive action commenced.

The whole problem consists of choosing the threshold to make sure that the number of V^k_{Ref} will decrease with the number of frames taken and does not converge to a number greater than 1. The convergence speed must also be optimized.

The mathematical choice of thresholds related to the choice of weights used to estimate the distances is a theoretical problem just as complex as that of the estimation of probability densities in the Bayesian classifier. They are directly related because they are merely restatements of the same problem.

This procedure is essentially a practical one and is therefore significant. In fact, it allows the operator to avoid sophisticated design and the need for linear calculations; a setting on the robot can be changed by adjusting a threshold voltage without forcing the operator to become involved with calculations.

The greatest difficulty arises when a number of distances are found in the vicinity of the threshold, and are therefore all capable of having attributed V to each of their V^k_{Ref} values – all those situated above the threshold are to be eliminated even if they are very near. So a method employing two thresholds can be used. The distances below the original threshold plus those above are retained, but at a distance from the threshold. Also all those above are related, but at a distance from the threshold smaller than a particular small value. An adaptive threshold could also be created. This could be done by drawing a curve that divides the distances into two groups – those that will be retained and those that will be rejected – based on the double criteria of the absolute value and of proximity.

8.6 The decision trees

In discussing the representation of an image in Chapter 7 possible methods were briefly summarized (most of which apply more to individual objects than to complex scenes). It could be assumed that matrix V^k_{Ref} is formed by the juxtaposition of these individual representations. Many would be either redundant or useless with respect to the problem presented to the robot. It should be remembered that the robot is asked the following questions: Which object is it? Where is it? What is its spatial orientation? To generate appropriate answers, recognition of the object must be made, which up to now has been obtained by recognizing similarity between measured representation and a reference representation.

Recognizing an object also involves distinguishing it from other objects ie eliminating them in a choice procedure. The method of recognition using a decision tree is based on attributing measurements to some objects and not others. The original set of objects is partitioned into successive categories, regrouping fewer and fewer objects. This method provides another type of classifier, used in a probabilistic, or more often, deterministic way.

Taking an example to demonstrate the essence of this technique, imagine that there are five objects: a cube, a sphere, a flat disk, a cylinder and a pyramid with a rectangular base. Suppose these objects are placed in a horizontal plane, that a single object is completely visible using a mobile camera initially situated vertically above the object. The shape of the possible outlines obtained when the camera is moved is given in Figure 8.10. There are 13 possible cases.

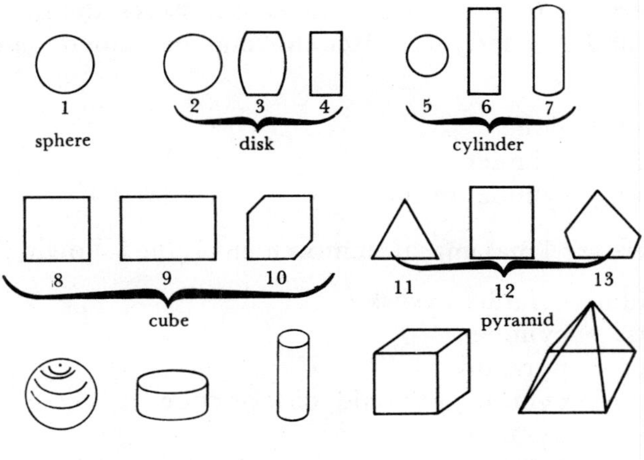

Figure 8.10. *Possible contours of five objects*

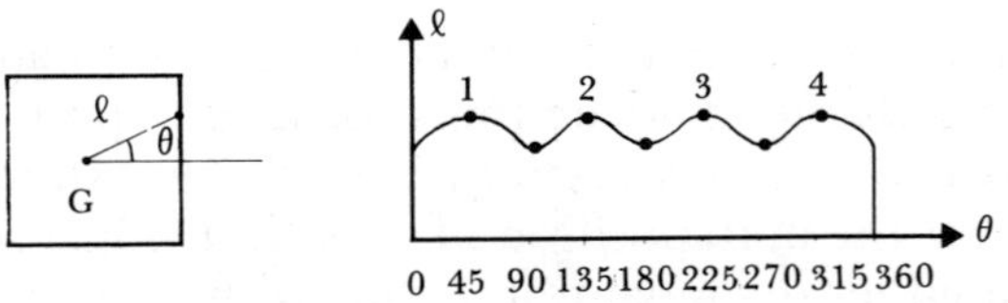

Figure 8.11. *Detection of maximum radii starting from G and ending at the edge. There are 4 maxima*

To recognize the objects, the relative maxima of radii from the object's centre of gravity to its outline (Figure 8.11) are chosen as representative.

Assume that the dimensions and the authorized angles of view are such that the maximum for each of the cases in Figure 8.10 is: *no 1*, 0; *no 2*, 0; *no 3*, 4; *no 4*, 4; *no 5*, 0; *no 6*, 4; *no 7*, 2; *no 8*, 4; *no 9*, 4; *no 10*, 0; *no 11*, 3; *no 12*, 4; *no 13*, 5.

Obtain the maximum numbers of maxima from above. The following criteria will be arrived at:

0 maximum:	sphere, disk or cylinder
2 maxima:	cylinder
3 maxima:	pyramid
4 maxima:	disk, cylinder, cube, pyramid
5 maxima:	pyramid
6 maxima:	cube

The indecision for 4 out of 6 possibilities has been removed.

Now tilt the camera to 30°, for example by directing it at the object, but only if there are 0 or 4 maxima.

When there are 0 maxima, the choice is between the sphere, the disk and the cylinder. Therefore, after moving the camera (movement 1) there are:

0 maxima:	globe
2 maxima:	cylinder
4 maxima:	cylinder or disk

When there are 4 maxima, after movement 1, the following is obtained:

0 maximum:	cylinder or disk
2 maxima:	cylinder
3 maxima:	pyramid
4 maxima:	cylinder, pyramid, disk or cube
5 maxima:	pyramid
6 maxima:	cube

Three ambiguous cases remain: where there have been successively

0 and 4 maxima, where there have been 4 maxima twice, where there have been 4, then 0 maxima.

Now consider moving the camera again to look at the object at an angle of 30° from the vertical, but in a plane perpendicular to the plane of movement. Calling this movement 2, the 0 maximum, then the 4 maxima case can only lead to 2 maxima if the object is the cylinder and 4 maxima if the object is the disk. The 4 maxima followed by the 4 maxima case can only lead to 2 maxima if with the cylinder, 3 with pyramid and 4 with the disk. Finally the case where there are 4 maxima first, then 0, will reveal after this second movement either 2 maxima for the cylinder or 4 for the disk.

This procedure forms a *decision tree* which leads to separation and recognition of the five objects using a simple, characteristic parameter

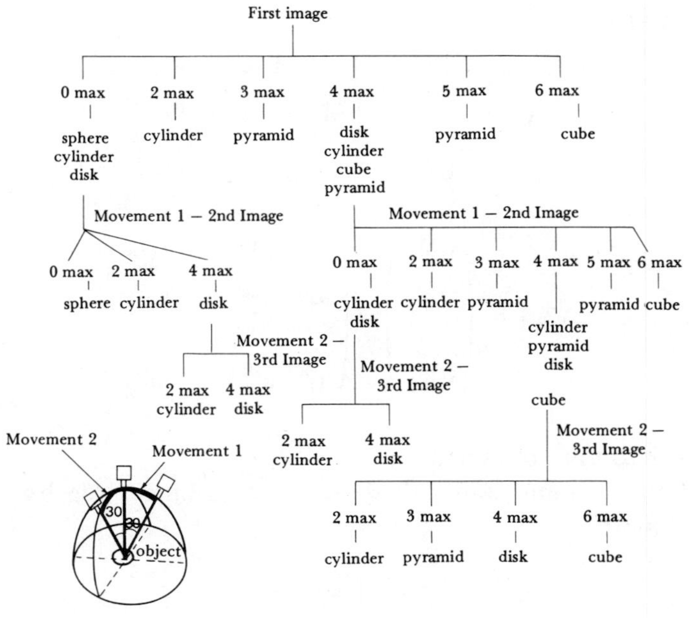

Figure 8.12. *Decision tree for recognition of three objects using three images and counting the maximum number of radii*

(Figure 8.12). Several things should be noted about this example:

1. Representation can be simplified, especially when the available information is increased (eg by taking frames from other angles).
2. Representation contains both 'descriptive' information and 'active' information. The first image tells what must be done next. It is in this sense that environment-robot interaction must be understood, using vision.

3. The validity of the decision tree is dependent on the quality of measurement. If the image produces noise there is only a certain probability of having 4 maxima in the image eg there will be orientation towards probabilistic decisions for this tree. However, the simpler the representation, the greater its precision, other things being equal. (Moreover, the fact that several frames can be taken increases the certainty of measurement.)

A decision tree can also be created using representation matrix V. This means that, instead of taking *the group* of characteristic vectors, balancing them and comparing the result with the V^k_{Ref} reference matrices by establishment of distances, this distance can be established for *each* characteristic vector. Combinations and comparisons of these distances allow a decision to be made using a tree. This case is particularly true when only one frame (2D) is taken.

Using again equations (8-14) and (8-15) for the measurement and reference matrices:

$$V = \begin{pmatrix} V^1 \\ V^2 \\ \\ V^\ell \end{pmatrix} = \begin{pmatrix} a_1^1 & a_2^1 & \dots & a_{n_1}^1 \\ a_1^2 & a_2^2 & \dots & a_{n_2}^2 \\ \\ a_1^\ell & a_2^\ell & \dots & a_{n_\ell}^\ell \end{pmatrix} \tag{8-14}$$

$$V^k_{Ref} = \begin{pmatrix} V^{k1}_{Ref} \\ V^{k2}_{Ref} \\ \\ V^{k\ell}_{Ref} \end{pmatrix} = \begin{pmatrix} a_1^{1k} & a_2^{1k} & \dots & a_{n_1}^{1k} \\ a_1^{2k} & a_2^{2k} & \dots & a_{n_2}^{2k} \\ \\ a_1^{\ell k} & a_2^{\ell k} & \dots & a_{n_\ell}^{\ell k} \end{pmatrix} \tag{8-15}$$

where $k = 1$ to t are obtained.

Then, taking a frame and calculating the distances (eg by the least weighted squares):

$$d_1^{k1} = \sum_{i=1}^{n_1} g_i^{1k} [(a_i^1)^2 - (a_i^{1k})^2] \tag{8-24}$$

$$d_2^{k2} = \sum_{i=1}^{n_2} g_i^{2k} [(a_i^2)^2 - (a_i^{2k})^2] \tag{8-25}$$

$$d_\ell^{k\ell} = \sum_{i=1}^{n_\ell} g_i^{\ell k} [(a_i^\ell)^2 - (a_i^{\ell k})^2] \tag{8-26}$$

k ranging from 1 to t.

A distance matrix of six columns and lines is obtained:

$$D = \begin{pmatrix} d_1^{11} & d_1^{21} & \dots & d_1^{t1} \\ d_2^{12} & d_2^{22} & \dots & d_2^{t2} \\ \\ d_\ell^{1\ell} & d_\ell^{2\ell} & \dots & d_\ell^{t\ell} \end{pmatrix} \qquad (8\text{-}27)$$

If a reliability analysis of each of these distances has been made beforehand, it enables a classification to be established, using the decision tree, based on: If d_1^{11} is less than a certain threshold and d_3^{35} greater than another threshold, and if d_2^{22} is close to d_6^{43} whilst remaining larger than d_3^{25}, one has such and such a property. So there is a series of logical equations comparable with those implied in Figure 8.12.

This method is interesting because, as with all methods based on the use of thresholds, recognition is quickly obtained (ie there are few calculations). The problem goes back to the design stage where relevance of the form of representation chosen may be judged, the decision tree established by discriminant, functional analysis or analysis of the principal components, beginning by taking test frames. On the one hand, every problem requires particular characteristic vectors, whereas on the other hand there are several possible decision trees, and finally, the variances of each distance being a function of the characteristic vector that produces it, the sub-optimal decision tree (the one that produces the highest success rate) is the one which involves minimum risk. That is to say that for different problems using the same characteristic vectors, the decision tree can be different if the same success rate is to be obtained.[122] This is a very important point in practice. There is an ordering of characteristic vectors.

8.7 Task-oriented features

When considering key features in an image, the eventual use must be considered. A distinction must be drawn between representation of the scene (the characteristic vectors) and key features, although they are linked and sometimes identical.

A distinction must also be made between key features used in *recognition* and those used in the *task*. The first often effectively coincides with the characteristic vector of the images. In robotics, three recognition features have to be considered: What objects are present in the scene? What are their positions? What are their orientations in space?

Consider an example in which a dumb-bell is to be lifted by the robot (Figure 8.13). The position and orientation of the dumb-bell are known and the dumb-bell is recognized using mask matching. This does not provide enough information for a successful operation because the

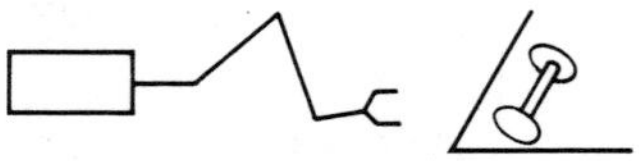

Figure 8.13.

diameter and weight of the dumb-bell are not known. Also, it is not known:

1. Whether the gripper, which supposedly has two fingers, could be opened sufficiently for the dumb-bell to be gripped.
2. Whether the weight of the object is excessive for the robot.

The two elements pertinent to the task are not supplied in the image recognition process. The first can be calculated using the recognition chosen, but not the second. The decision processes required are schematized in Figure 8.2 and again in Figure 8.14.

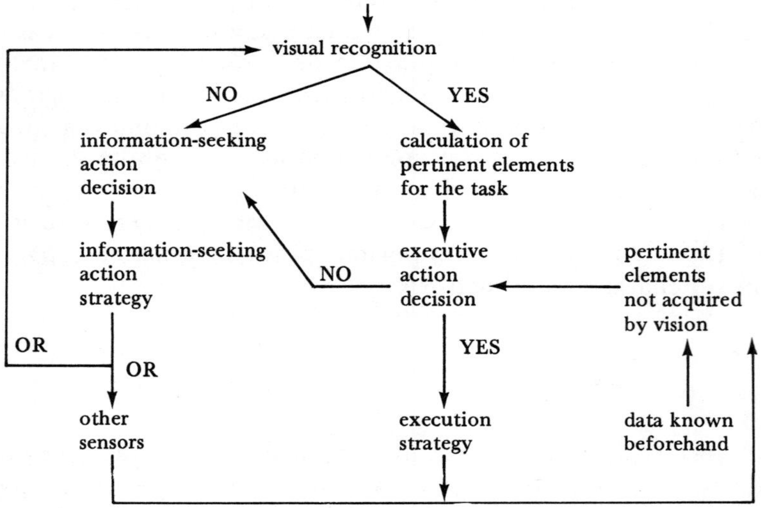

Figure 8.14. *Pertinent elements for the task and decision levels*

8.8 Optical illusions and image comprehension by a robot

Optical illusions fall into two main categories: 'static' illusions where a pattern appears such that the observer commits a recognition error and 'dynamic' illusions where, when observing moving objects, distortions appear, because of the sampling period of the human eye, retinal persistence and non-linearity of sensitivity of the retina.

Considering static illusions, two categories can be distinguished:

1. Those due to defective vision (ie long or short sight) or due to poor lighting conditions. In recalling a scene that has been witnessed, trying to interpret it with too few characteristic elements available leads to mistakes.
2. Those produced with correct vision, but with erroneous interpretation that exemplifies certain relationships between lengths, points on surfaces etc, or the presence of non-delineated figures.

The robot suffers from optical, static illusions of the first kind. It is always weak in interpretation because it lacks knowledge of characteristic vectors and precision in measuring the components of these vectors.

Static illusions of the second kind (for which the term 'optical illusion' is often wrongly used) are never normally present in the robot, but can be programmed in. Take an example drawn from Gregory[48] (Figure 8.15).

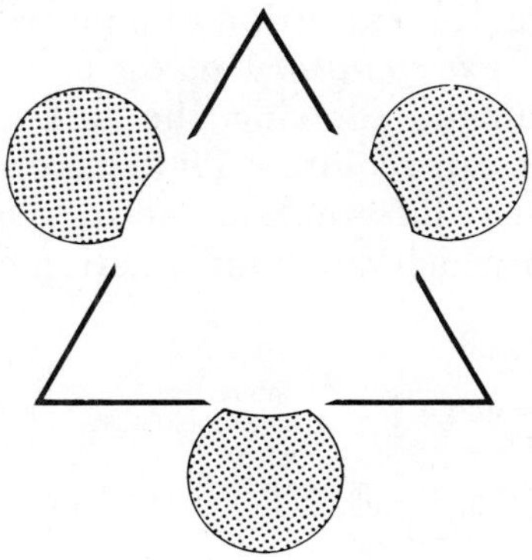

Figure 8.15. *Is there a white circle in the figure?*

In this case the optical illusion would be manifested either when:

1. The human being sees a white circle that is not present.
2. Or he 'sees' the white circle masking the object. The reasons for this phenomenon will not be considered here – it involves very complex issues of perception.

Could a robot 'see' these two same features?

1. Imagine that the robot has been asked to find the circles in this image using an algorithm which systematically forms a circle as soon as 25 points situated on the outline have been found, so that a circle may be drawn through them. When the image is quantized correctly and the outlines drawn, the robot detects the three black circles and the white circle.

2. On the other hand, it will be strictly incapable of deducing from it that the white circle hides the figure, unless the robot has been programmed to make this interpretation.

Point 1. is an effect connected with the lack of information of the characteristic vector. It is a simple error of interpretation.

Point 2. cannot be 'imagined' by the robot – it does not possess an imagination, only programs.

8.9 Distribution of information

The execution of a task or a macrotask by a robot which is interactive with its environment will be carried into effect by a series of interrelated sequences: deduction of information leads to execution of a micro-task or deduction of 'complete' information followed by execution of the macrotask.

In Chapter 1, Figure 1.4, execution of a macrotask was resolved into two principal phases: a *preparatory phase* in which the initialization information was obtained, allowing the task to commence and an *execution phase* which can require on-line information. The temporary development of pertinent information can be represented symbolically as in Figure 8.16, from which much information can be drawn.

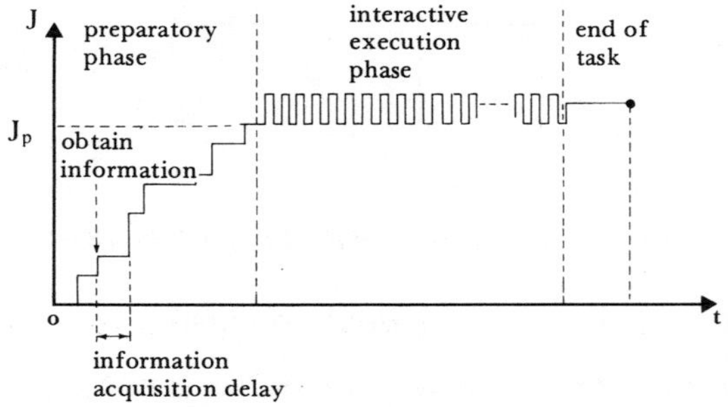

Figure 8.16. *Figurative representation of evolution of pertinent information*

1. To commence execution of the task, a certain quantity of information J_p is needed. There are two possible sources: prior information J_a and information J_c derived from the sensory organ. So

$$J_p = J_a + J_c$$

With J_p constant, the more prior information we have, the less is needed from the sensors. On the other hand, the less the robot will be capable of carrying out diverse tasks, because prior data represent an environmental constraint.

2. Note that in the preparatory phase, the information grows in steps of unequal duration. The amplitude step represents the symbolic jump made when a certain quantity of pertinent information has just been acquired; the duration represents the acquisition time which can imply manoeuvres of the robot and exploitations of images or other signals.

3. In the interactive phase of execution, it is assumed that the information obtained in the preparatory phase is held in the memory and remains pertinent. The pertinent information used in completing the task has, on the other hand, a *transitory* nature. Suppose a movement is carried out whilst checking a force between robot and environment. At time t the pertinent information is derived from the difference between the value desired and the value measured, but at the next moment these values are no longer pertinent – the new values needed and measured become pertinent.

So the information needed for a robot is characterized by:

(a) The fact that it can be provided (prior information) or acquired (before or during an execution of the task).
(b) The fact that this information is by nature partly permanent and partly transitory.

4. The stepwise evolution of the growth of information in the preparatory phase deserves more attention. To make a realistic representation on a curve a mathematical definition of what pertinence of information is, would first be needed. It was seen in equation (8-7) what was pertinent for a robot which executes a task. Now, an attempt will be made to obtain a measurement of pertinence.

Suppose that, to deal with the execution of the task, three pertinent values (a_1, a_2, a_3), assumed to be independent, are acquired and in content need to be known.

Having three values, the task can be carried out, but there are disturbing elements that might prevent satisfactory completion.

The probability of success is written as:

$$P(\text{success } a_1 \; a_2 \; a_3)$$

The pertinence of, for example, a_1 may be assessed in two ways:

– Either by assessing probability of success of the task, knowing (having acquired) a_1 and taking for a_2 and a_3 values at random, a_2^* and a_3^* in a finite, discrete or continuous set, supplied as prior information (eg if a_2^* represents a length, and knowing that in the problem presented there are only lengths between 2 and 6 cm, a length will be chosen at random in this group). This probability will be called $P(\text{success } a_1 \; a_2^* \; a_3^*)$
– Or, on the contrary, using probability of failure, having acquired a_2 and a_3, and taking a_1 at random on an authorized finite group:

$$P(\text{failure } a_1^* \ a_2 \ a_3) = 1 - P(\text{success } a_1^* \ a_2 \ a_3) \qquad (8\text{-}29)$$

The first probability shows what knowledge of a_1 alone brings, the second the 'lack of winning' resulting from its absence.

So the pertinence of a_1 can be estimated using:

$$\text{Loss } (a_1) = ½(P(\text{success } a_1 \ a_2^* \ a_3^*) + P(\text{failure } a_1^* \ a_2 \ a_3) \qquad (8\text{-}30)$$

$$= \frac{1 + P((\text{success } a_1 \ a_2^* \ a_3^*) - P(\text{success } a_1^* \ a_2 \ a_3)}{2} \qquad (8\text{-}31)$$

Example: Suppose that from the prior information it is known that: a_1 can take any continuous value between a_{1min} and a_{1max}; a_2 can take any discrete value from q values; a_3 can take any continuous value between a_{3min} and a_{3max}.

Suppose, on the other hand, that the real values of a_1, a_2 and a_3 are a_1^{des} a_2^{des} a_3^{des} and that if a_1, a_2 and a_3 are measured, values such as:

$$a_1^{des} - \Delta a_1^{des} \leqslant a_1^{mes} \leqslant a_1^{des} + \Delta a_1^{des} \qquad (8\text{-}32)$$

$$a_3^{des} - \Delta a_3^{des} \leqslant a_3^{mes} \leqslant a_3^{des} + \Delta a_3^{des} \qquad (8\text{-}33)$$

are found, where a^{des} is the desired value of a; and a^{mes} is the measured value. $a_2^{mes} = a_2^{des}$ or one of the nearest K values, then:

$$P(\text{success} \mid a_1 \ a_2 \ a_3) = P_a \qquad (8\text{-}34)$$

So:

$$P(\text{success} \mid a_1 \ a_2^* \ a_3^*) = P(a_3^{des} - \Delta a_3^{des} \leqslant a_3^* \leqslant a_3^{des} + \Delta a_3^{des})$$
$$xP(a_2^* = a_2^{des} \text{ or one of the nearest})xP_a \qquad (8\text{-}35)$$

$$P(\text{success} \mid a_1 \ a_2^* \ a_3^*) = \frac{2\Delta a_3^{des}}{(a_{3max} - a_{3min})} \cdot \frac{(2K+1)P_a}{q} \qquad (8\text{-}36)$$

in supposing that the choice of a value of a_2 and a_3 has a constant probability.

Likewise:

$$P(\text{success} \mid a_1^* \ a_2 \ a_3) = P(a_1^{des} - \Delta a_1^{des} \leqslant a_1^* \leqslant a_1^{des} + \Delta a_1^{des}) \cdot P_a \qquad (8\text{-}37)$$

suppose:

$$P(\text{success} \mid a_1^* \ a_2 \ a_3) = \frac{2\Delta a_1}{a_1^{max} - a_1^{min}} \cdot P_a \qquad (8\text{-}38)$$

So the pertinence of element a_1 is:

$$\text{Pert}(a_1) = 1/2 + P_a \left\{ \frac{(2K+1) \cdot |\Delta a_3^{des}|}{q|(a_{3max} - a_{3min})|} - \frac{|\Delta a_1^{des}|}{|(a_{1max} - a_{1min})|} \right\} \qquad (8\text{-}39)$$

This pertinence is a probability and cannot exceed 1 or, more exactly, $(1 + P_a)/2$.

One could equally take as characteristic the negative entropy of this pertinence:

$$H[\text{pert}\,(a_1)] = -\,\text{Pert}\,(a_1) \cdot \log \text{Pert}\,(a_1) \qquad \textbf{(8-40)}$$

The extension to any number of pertinent features is immediate. The pertinence of any known number of elements could be found. In the previous example:

$$\text{Pert}\,(a_1, a_2) = [1 + P(\text{success} \mid a_1\ a_2\ a_3^*) - P(\text{success} \mid a_1^*\ a_2^*\ a_3^*)]\,/2 \qquad \textbf{(8-41)}$$

which is not the same thing. It is important to note that

$$\text{Pert}\,(a_3) = [1 + P(\text{success} \mid a_1^*\ a_2^*\ a_3) - P(\text{success} \mid a_1\ a_2\ a_3^*)]\,/2 \qquad \textbf{(8-42)}$$

If elements $a_1\ a_2\ a_3$ are dependent, the pertinence can be formulated if the dependence is known. Return to the previous example and suppose that a_2 and a_3 have a dependence with respect to a_1, but not the other way round. This means that, if a_1 is measured, a_2 and a_3 will be constrained to a group of narrower values than before. For example: a_2 can take any discrete q' value if a_1 is measured with $q' < q$ (and $2K + 1 < q'$). a_3 can take any value between a'_{3max} and a'_{3min}, if a_1 is measured, with a'_{3max} and a'_{3min} within interval (a_{3max}, a_{3min}).

On the other hand, nothing has changed for the values that a_1 can take if a_2 and a_3 are measured. The new pertinence for a_1 will be:

$$\text{Pert}'\,(a_1) = 1/2 + P_a \left[\frac{(2k+1)}{q'} \quad \frac{\Delta a_3^{des}}{(a'_{3max} - a'_{3min})} \quad \frac{-\,\Delta a_1^{des}}{(a_{1max} - a_{1min})} \right] \qquad \textbf{(8-43)}$$

and:

$$\text{Pert}'\,(a_1) > \text{Pert}\,(a_1) \qquad \textbf{(8-44)}$$

In Figure 8.16 every value in the preparatory phase can be interchanged with the pertinence of the acquired element.

If the process is limited to recognition, the preceding problem can be formulated by identification of the pertinence of a parameter with mutual information.[123]

Consider a series of objects to be identified using a series of frames. Taking an object at random, recognition is made by use of a group of characteristic parameters $(\alpha_1, \alpha_2 \ldots \alpha_q)$. Each parameter α_j can take every value within interval (x_j^{in}, x_j^{fi}) with a certain law of probability $p_j(x_j)$. Recognition consists of attributing the object, after measurement of the characteristic parameters, to class C_k.

How is the parameter found which, in a frame, will provide maximum information ie will be the most pertinent?

It is the parameter that will maximize the mutual information

$$I\,(\Omega, P_j)$$

Ω defining the situation: the object belongs to class C_k; P_j defining the situation: the jth parameter has a value which belongs to the ℓth interval of interval $(x_j^{in}\ x_j^{fi})$ that has been *discretized.*

Consider two random variables X and Y, X having K *modalities* $X_1 \ldots X_k$ and Y having p modalities $Y_1 \ldots Y_p$. The mutual information between two events X_m and Y_n is given by:

$$I(X_m, Y_n) = \frac{\text{Log}\,[p(X_m/Y_n)]}{p(X_m)} = I(Y_n/X_m)$$

The mathematical expectation on X of the mutual information defines the useful information brought by event Y_n on situation X:

$$I(X, Y_n) = \sum_m p(X_m \mid Y_n) \frac{\text{Log}\,p(X_m \mid Y_n)}{p(X_m)}$$

and the average mutual information between the two random variables X and Y can be written as:

$$I(X, Y) = \sum_n p(Y_n)\, I(X, Y_n)$$

$$I(X, Y) = \sum_m \sum_n p(X_m, Y_n) \frac{\text{Log}\,p(X_m \mid Y_n)}{p(X_m)}$$

The random variables now being Ω and P_j, $I(\Omega, P_j)$ can be considered a value characterizing the information introduced by the jth parameter to the recognition. This is the pertinence of that parameter; it can, for example, be used to choose a hierarchy of parameters to be used for each frame.

8.10 Returning to discriminant functions

It has been seen that a discriminant function allows for action decisions to be established. In the Bayesian classifier it was identified with the function that allows the choice of least risk and it implies a probability. The introduction of these probabilities is connected with high possibility of error in measuring representation of the images. If simple representations and decisions tolerant of measurement error can be found, deterministic, discriminant functions which are also very simple will be possible. This, of course, supposes that the complexity of deciphering a single frame is replaced by the simplicity of information concerning several frames. This is a choice that already has been set out:

pertinent information with *one* frame + complex algorithm	$\cong$	pertinent information with a number of frames + simple algorithm

For example a robot equipped with a camera on its end effector has to carry out the following task: to grasp within its reachable space, a cylinder of diameter such that gripping is possible, but of unknown height (more than 10 cm). The robot has to grasp the cylinder at its middle. The cylinder can be upright or recumbent in a known, horizontal plane and is present alone in the robot's environment.

Rather than photographing the environment and searching for the pertinent elements by complex image analysis, the mobility of the camera and the environment-robot interaction can be used by limiting the process to counting points as a discriminant function. The scenario can be as follows:

Task 1: Place the robot in the initial position. This movement is made using a control in open loop with respect to the environment.

Task 2: Search for the cylinder by scanning by camera the workplane, with a double rotational movement about the vertical and about a mobile horizontal axis (Figure 8.17). For every movement, the number of luminous points on the optic sensor is measured (it is assumed that the cylinder is white and the work black, or the other way round). If this number exceeds a given threshold (the discriminant function), the

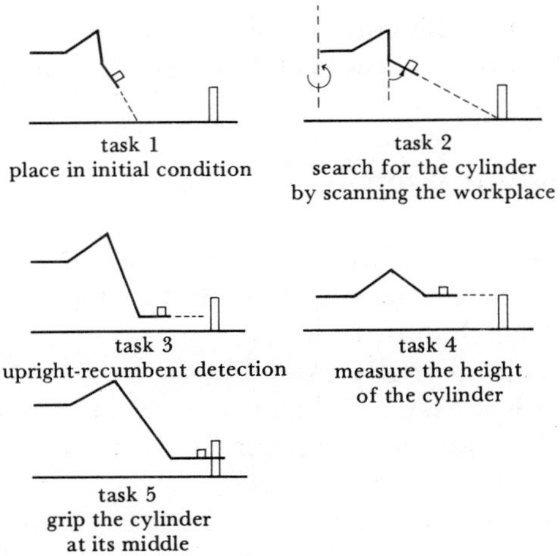

Figure 8.17. *Strategy for gripping a cylinder using discriminant functions in the image. It is reduced to counting a number of luminous points, and produces a high tolerance of errors*

process is stopped; either the base of the cylinder is found, if it is upright or any zone if it is recumbent.

Task 3: The robot is moved so that the camera looks horizontally at the vertical of the zone previously detected, but 10 cm above this zone. There again, the discriminant function is a simple threshold comprising a number of luminous points. If this number exceeds the threshold, the cylinder is upright, otherwise it is recumbent.

Task 4: If it is upright, the vertical of the zone detected in Task 2 will continue to be inspected, until the number of points is zero. Thus the height of the cylinder is found. (The discriminant function remaining simple.)

Task 5: Calculate the coordinates of the centre of the cylinder.

Task 6: Control the gripping.

Task 4 again: If the cylinder is recumbent, the zone detected in Task 2 is looked at from above and a similarly simple strategy, based on discrimination by counting the number of luminous points present on the optic sensor, is used to find the orientation and outer ends of the cylinder.

This example illustrates that, accepting environmental constraints and using a mobile sensor, the discriminant functions become simple and very tolerant of errors, and the success rate is not reduced by the sensitivity or complexity of the processing of the images. Note, in passing, that visual information contains both descriptive information about the object (reduced to presence-absence) and active information, because the result of the observation indicates, moreover, where to seek earlier information. Finally, if the visual success rate is 100 per cent (complex image analysis would reduce this rate), one exchanges recognition time for the need to move the robot. This is an illustration of the standard precision-speed dilemma or of another, more general dilemma: measurement affects task completion, because it prolongs its duration.

8.11 Cameras for industrial robots

In assessing the available methods of robot vision it can be seen that, as robotics is a recent phenomenon, users have employed commercially available sensors rather than others.

There are some commercially available devices: system MIC, system AUTOMATICS (with a General Electric camera and an LSI 11) and laboratory devices (eg outline extractor Systems Analysis at the Automatics Laboratory at Toulouse).

Which criteria should be used to construct a camera of this type? Examination of this problem will be restricted to a 2D camera usable for K-2D. The camera to be considered is not intended for inspection purposes, but for cooperation with the robot, which has to manipulate objects.

First of all, to make a 'universal' camera, artificial vision systems that are as good as the human system would have to be created.[124] This is plainly not practical, so the objective is to produce systems that can be of use in the solution of practical problems. This is possible:

1. If designed for the individual static robot, equipped with one or more fixed or mobile cameras.
2. If one can confine applications to:
   - The recognition of objects.
   - The recognition of positions of objects.
   - The recognition of spatial orientation of objects.
3. Or if one can apply realistic but very simplified constraints:
   - No overlapping of images of objects.
   - An object is always seen in its entirety.

These restrictions are, in actual fact, very reasonable and leave a large field of applications.

In these conditions, the robot camera will be distinguished from the conventional camera in that it will not supply an image composed of points, but a group of characteristic vectors – a matrix such as V in equations (8-12) to (8-14). So characteristic vectors will be produced, based on the geometry, topology or functions of points of the outline of the objects.

How is the set of extractors of characteristic vectors chosen? Several elements should be taken into consideration:

1. A segmentation stage separating the objects in the image, counting them and determining the position of their centre of gravity, is imperative.
2. The resulting characteristic vectors will relate to each object.
3. Their number must be limited for reasons of cost – in terms of camera and calculation time. In fact, many problems involving a functional, characteristic vector such as the three or four first Dudani moments, or three or four geometrical vectors (surface, form factor, minimum and maximum radii etc) and some topological data, can be resolved.

Technologically this camera must be small, so that it can eventually be integrated with the robot's structure. CCD sensors respond well to these constraints. On the other hand, definitions in the region of 10,000 to 40,000 points can be achieved without changing the success rate of execution of a task by a robot with vision. This not only coincides with the technology of the CCD, but is, moreover, a favourable factor for the speed of extraction of characteristic vectors. As these are calculated on a binary image, stages of digitization, extraction of outlines and of thresholding can follow the analog output of the CCD matrix.

The camera can be used in two ways. A totally heuristic analysis indicates that there is no need for some of the available characteristic vectors (eg if there is only one object, the segmentation stage and the topological vectors are, in general, not required). If the camera is programmable it can be used in temporary applications.

To exploit the remaining vectors, one can again proceed heuristically or by tests and trials, but preferably by simulation. For, in the robot, recognition should be made in a few dozen milliseconds. This involves use of a minimum number of characteristic vectors in a sub-optimal decision tree.

A simulation program can be designed in the following way (Figure 8.18). Beginning with images of each object the selected characteristic vectors are calculated, and their stability estimated with respect to the quantization of the image and parallax.

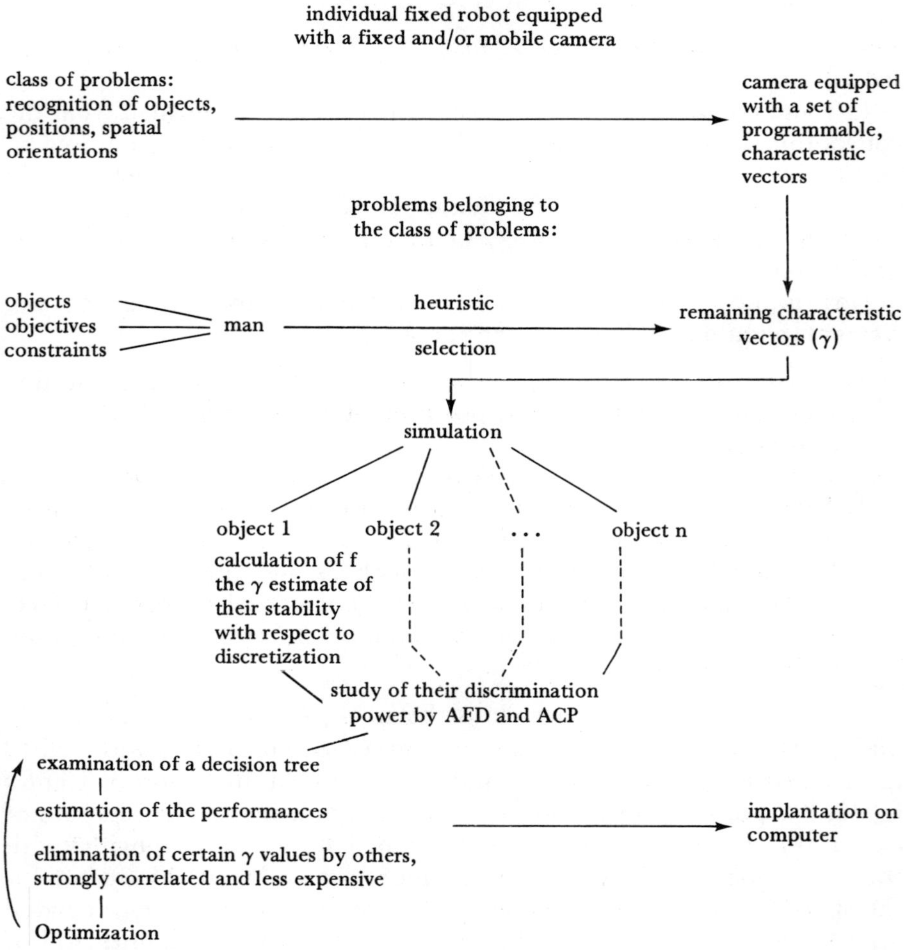

Figure 8.18. *Principle of use of a camera delivering a programmable set of characteristic vectors*

The power of discrimination of these characteristic vectors is tested for the group of objects involved eg by discriminant functional analysis or by analysis in principal components.

After that, a decision tree is considered and the performances tested. By studying the correlation, a number of vectors that are strongly correlated with other, more costly ones, can be eliminated. The decision tree is then optimized and implanted on the computer connected with the robot.

However, for a robot equipped with a system of vision, the discriminant functional analysis possessess two defects:

1. In principle, clusters of points — representing each object — are separated, using optimizing criteria. As a result, the zones of uncertainty are eliminated. The decision tree deduced from this inevitably leads to recognition. This can be misleading if the groups of points corresponding with two distinct objects are insufficiently retrieved. To obtain certainty of decision, the decision tree would need to be able to respond with indecision between two classes as a legitimate solution to a problem. With the AFD this does not seem possible.
2. The technique used in establishing a learning program that remains unchanged on-line. In fact, the introduction of a new object requires the introduction of the cluster (ie groups of points) associated with it ie a number of frames of this object (by varying parameters such as the deformation due to parallax and the orientation of the object with respect to the grid on the sensitive surface), then the generation of new hyperplanes partitioning the space of the parameters and, finally, the generation of the new decision tree. These procedures are complex and can be protracted.

One means of coping with these defects consists of estimating a distance between the clouds, so that there is then a possibility of thresholding distances, permitting variations in the decision tree. (On the other hand, the possibility of estimating the correlations between parameters present in the AFD is lost.) Regulating the thresholds can be delicate. It provides a possibility of a choice between several solutions at the end of the decision tree.

Finally, knowing the probability distribution of the various parameters involved, the Bayesian classifier associated with supervised learning can provide excellent results. This distribution seems at present best designed using the probability histogram connected with the parameters.

Be that as it may, the robot camera must be characterized by three elements:

1. A 'physical' system that supplies, on-line with the output of the camera, a number of characteristic parameters (simple to select for a given category of problems).

2. An interface card with a common digital computer.
3. A learning program easy to implement on the computer. This learning program (supervised or not) must have certainty of decision and a recognition capability offering a choice between one object and another. This allows the user to cancel this last choice by using a complementary system (eg another camera).

Such complete cameras can be built at an acceptable cost.

Conclusions

The two principal constraints that limit exploitation of images for robots interacting with their environment are time and reliability of the decisions involved. For industrial robots two others may be added: cost and reliability of the hardware. Let us quickly review these four factors.

1. Time: When a robot is ordered to execute a task using a mobile camera, the execution can be divided into two successive phases: acquisition of pertinent information by K frames and execution of the task, supposed to be permanently supervised by a camera. The process will take place in the following steps:

– Setting up the system	duration: t_0
– First frame Extraction of pertinent elements First decision	duration: t_{11}
– Generation of the camera's first movement Execution of the first movement	duration: t_{12}
– Second frame Extraction of pertinent elements Second decision	duration: t_{21}
– Generation of the second movement Execution of the second movement etc . . . until	duration: t_{22}
– Kth frame Extraction of pertinent elements Kth decision: start of execution of the task	duration: t_{K1}

The second phase will be identical in principle, but will generate small movements of the robot and be supervised using an image.

Thus there will be K′ durations Δt_{i1} and Δt_{i2}. And the duration of the task will be:

$$T = t_0 + \sum_{j=1}^{K} t_{j1} + t_{j2} + \sum_{i=1}^{K'} \Delta t_{i1} + \Delta t_{i2} \qquad \text{(8-45)}$$

This time is divided into logical time T (calculation time), and a time T_r corresponding to the time taken for movements of the camera and the robot. T_r cannot be changed (Volume 1).

The calculation time is an element which one can and must influence. For processing the image, the use of wired logic must be considered as thoroughly as possible. But there comes a point where wiring is no longer possible for reasons of cost, volume or complexity. Microprogramming and then programming must be used instead.

Parallel working of the calculations or time sharing of calculation and movement (work in concealed time) must, however, be exploited.

Finally, and this is not easy, the methods of processing images and the generation of trajectories must be carefully chosen. In the previous example, it may be possible, and better, in the information phase, to take a photograph without moving the camera, and to extract more characteristic vectors.

At present, work is proceeding towards use of such multiprocessor structures[125–128] or more precisely, multimicroprocessors,[129,130] which also can allow the robot with vision to attain speeds compatible with industrial needs.

2. Certainty of decision: The ultimate industrial robot would need to attain success rates close to those of man. As soon as vision is used with a robot, this success rate decreases greatly. To remedy this, the following steps might be taken:

1. Improve technology to have less noisy images which are also more tolerant of lighting variations. The improvement in sensor definition, if it can improve the precision of the characteristic vectors, is equally important to augment the speed of processing of the images.
2. Use the redundancy of information using multiple views. But there again, if the same camera is used, there is a penalty as a result of time delay.
3. Improve the decision strategy to profit from informational redundancy, which is often latent.
4. Find better characteristic vectors and discriminant functions tolerant of errors.
5. For each specific problem, choose specific devices available.

3. Cost: If a vision system is connected to a robot, the price must support comparisons with the price of the robot itself. The price of CCD or television cameras is reasonable, but they must be drastically redesigned to include processors, which are expensive to develop.

4. Reliability: At present, vision systems are being transferred from the research laboratory to commercial industry. However, there are many steps still to be taken to prove their industrial viability.

Chapter 9

Vision systems for industrial robots: a review

Artificial vision in robots is the subject of much research. Although at the time of writing few projects have advanced to the industrial stage, there are many experimental projects being undertaken, a few examples of which will be quoted. The first part of this chapter is devoted to work in the two-dimensional domain. The second part deals with work on depth information.

9.1 Work illustrating the two-dimensional technique

9.1.1 OBJECTIVES OF THE AUTHORS

Agin,[131] at the Stanford Research Institute, has worked on inspection and assembly of mechanical parts by robots with vision. The aim was the inspection of factory parts on a conveyor belt. The work consisted of either part recognition or quality control. Agin's[132] objective was the assembly of mechanical parts; the manipulation had as its aim the location of the position of a screw-hole followed by the placing of the screws in position.

Chien and Jones[133] at the University of Illinois carried out experiments on recognition and on the pursuit and grasping of moving objects. Several applications were attempted:

- *Supervision*: a camera gives a view of a street; the moving objects were followed and identified (cars, vans, bicycles etc).
- *Arrangement*: cubes placed on a table were found, then piled up one by one.
- *Assembly*: the robot had to assemble two parts. One was held by the robot. The other was held by a man who tried to prevent assembly by moving the part on the table at a speed compatible with that of the robot.

Hale and Saraga[134] at the Mullard Research Laboratory used a vision system to determine the location of the holes to be made in printed circuits, and to control the movement of the drill.

Kashioka *et al.*[135] at Hitachi carried out an experiment involving two anthropomorphic manipulator arms, a visual system and stress feedback.

The assembly of the parts of a vacuum cleaner was its object. This involved recognizing and finding those parts which could be grasped, then transporting them to the assembly zone and verifying their positions for assembly.

Lubinski *et al.*[136] at Virginia Polytechnic Institute and State University studied a system that supervised the position of an object and its movements in a given area.

Niemi *et al.*[137] at Helsinki University were also interested in recognition, the pursuit and grasping of objects. The recognition work consisted of classifying cylinders of different heights, lying on the work surface, and placing them in boxes according to their dimensions. The pursuit and grasping of moving objects was carried out on cylinders or spheres placed on a sloping plane. The trajectory needed to recapture them was calculated.

Perkins[138] of the Research Laboratories at General Motors worked on the recognition of objects of complex shape. This work was concerned with the determination of the position and orientation of (isolated) mechanical parts, under conditions of high noise level in the image.

At Olivetti's Industrial Automation Systems, Salmon[139] studied problems of representation, recognition and orientation of flat mechanical parts with a view to assembly.

Tani *et al.*[140] in Japan carried out an experiment involving the assembly of a rectangular-shaped part into a hole of the same shape with very little play, using only *visual* feedback.

Tsuji and Nakamura[141] and Yachida and Tsuji[142,143] at Osaka University carried out experiments on parts of motorcycle engines ie on recognition, position and orientation required to carry out *fitting*.

9.1.2 LEARNING PROCESSES

When experiments are to be repeated on the same objects, the system may be pre-programmed by the designer.

Agin suggests a dialogue between the system and the operator in these circumstances. A viewing bay was used to monitor the development of the process and the method of acquiring a new object. The program indicated the characteristics to be stored and was organized in a modular way, so that the algorithms with which characteristic vectors are exhausted were easily modified.

Yachida and Tsuji used a similar process. A program allowed the model of a new object to be generated in an interactive way, by presenting it to the camera. Thus the representation of an object could be modified. For example, for an object presented a characteristic vector can be derived. The system indicated whether this vector had already been used for objects already stored or was close to them. In such a case, it was regarded as providing no new information, and another one was tried.

9.1.3 METHODS OF IMAGE ACQUISITION

Agin[131] carried out tests using a television-type camera sampled on 120 × 120 points and digitized on 5 bits (32 grey levels). The image was then binarized with respect to a predetermined threshold. In later work this camera was replaced by a small array of 128 photodiodes placed above the conveyor belt. Using the camera, a complete image was stored and then processed line by line. With the small array the processing was carried out for each new line obtained by a movement of 1.2 mm of the conveyor belt and only the useful information present in the line was kept. Lighting was arranged in such a way that the image did not necessitate any filtering (Figure 9.1).

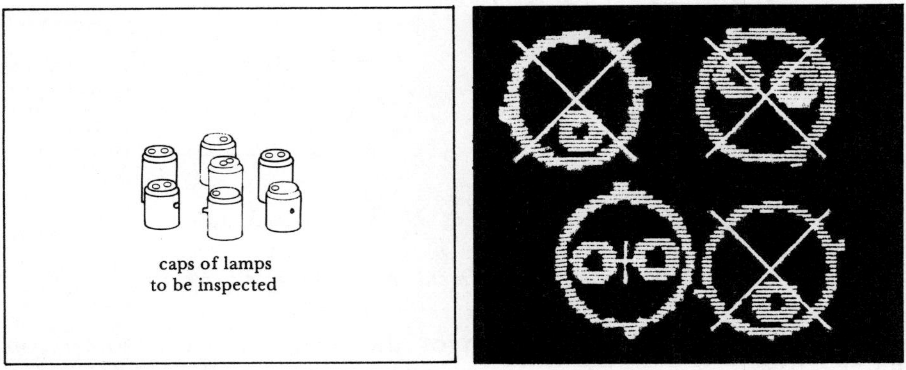

Figure 9.1. *The Agin inspection experiment*

Agin[132] studied several camera positions and their possible location on the manipulator. The feeler was formed by a CCD matrix of 100 × 100 points definition. The analog signal was thresholded directly, which led to serious constraints on the lighting. The imaging was done at predetermined locations (Figure 9.2).

Chien used a TV camera sampling 252 × 252 points. The camera was placed vertically and a mirror provided a lateral view of the scene.

Hale used a TV camera positioned horizontally with a mirror placed at 45°, providing vertical vision. A cross line provided human verification of the centring of the board (Figure 9.3).

Kashioka used eight TV cameras. Three served to determine the position of the vacuum cleaner parts and their graspability. Five were applied to inspection at the moment of assembly, one of which was connected with the robot's end effector using a system employing optic fibres

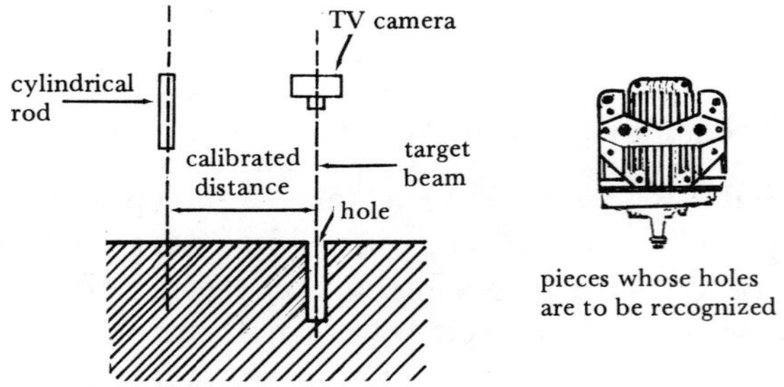

Figure 9.2. *The Agin assembly experiment*

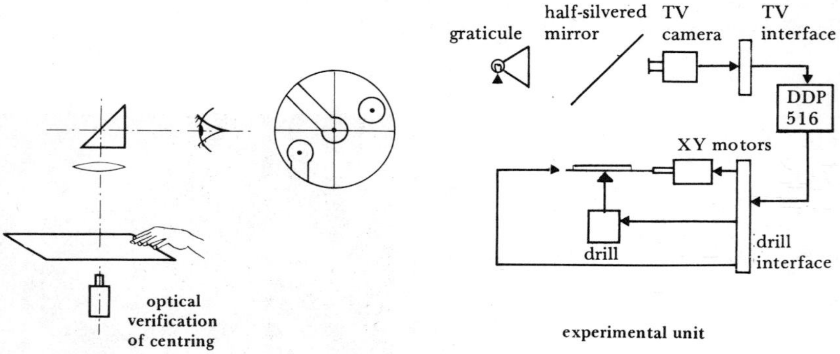

Figure 9.3. *The Hale experiment: piercing printed circuits*

(Figure 9.4). Spatial quantization of the images was carried out on 256 × 256 points and the video signal binarized with the help of a variable threshold as a function of the luminosity of chosen parts of pieces, global lighting being normalized. The outlines were extracted by spatial differentiation and polishing.

Lubinski placed a camera above the zone to be supervised. No attempt was made to recognize the object but simply to find out its position, and the signal was not digitized. The analysis consisted of triggering the two counters X and Y at the start of scanning (Figure 9.5). For every video line, the edge of the image served as a guide-mark to reset counter X. The presence of the object was characterized by a fall in the output level of the video signal below a predetermined threshold; this served to stop the counters, thus indicating the coordinates of the object.

Niemi used a CCD camera with a resolution of 50 × 50 placed above the workplane of the manipulator, allowing registration of an image in five milliseconds. The image was binarized with respect to a fixed threshold. The lighting was constrained (Figure 9.6).

Perkins[138] used a TV camera sampling 256 × 256 points digitized to 32 grey levels. The lighting was controlled in the learning phase, but varied slightly during manipulation (Figure 9.7).

The optic sensor used by Salmon consisted of a unique photoelectric cell receiving the light produced by a pencil of rays of 3 mm^2 cross-section incident on a revolving mirror (Figure 9.8). An area of 200 m^2 was thus scanned in 500 milliseconds. The image was divided into 200 × 200 points and binarized by fixed thresholding. It was filtered by smoothing.

Tani laid out three TV cameras, of which two were linked with the end effector of the manipulator using optic fibres; the third was positioned below the workplane. The images were sampled using 256 × 300 points, but windows supplied by the camera could be selected. These windows were divided into 256 × 300 equal points, thus improving the definition of the image. A binary image was formed by thresholding, requiring optimal lighting conditions (Figure 9.9).

A TV camera that sampled 128 × 128 points and digitized 64 grey levels was used by Yachida and Tsuji (Figure 9.10). The threshold was chosen from a histogram of the grey levels.

9.1.4 REPRESENTATION OF THE IMAGES

Agin began with a topological representation of the image, following connectedness analysis. Regions of equal grey level were separated, which allowed isolated objects and holes to be located. Extraction of the outlines was achieved using an operator of the Sobel's or Roberts' type. On these outlines the definitive representation was calculated. It comprised: the surface area, the perimeter, the square root of the surface area, the number and positions of the centre of gravity of the holes, the maximum, minimum and average values of the vector radius going from the centre of the image to the perimeter, the minimal circumscribed rectangle, central moments of first and second order.

For Hal's experiment, the image of the plate was represented by the number of holes to be pierced, matched by their coordinates and their diameters. The analysis of a hole took into account a 'roundness' factor.

Like Agin, Hashioka divided the image into regions using the grey levels. Niemi spatially differentiated the image using a 3 × 3 window and extracted the centre of gravity and the cylinder's orientation (the experiment was only concerned with cylinders).

Perkins examined the outlines using the 32 grey levels, and applying Hueckel's operator (section 7.3.1.3). Detected points were thus classified using categories of equal intensity and proximity, and smoothed using straight lines or arcs of circles based on changes of curvature.

Salmon used polar representation to describe objects. Tani approximated the outlines using segments of straight lines obtained with the

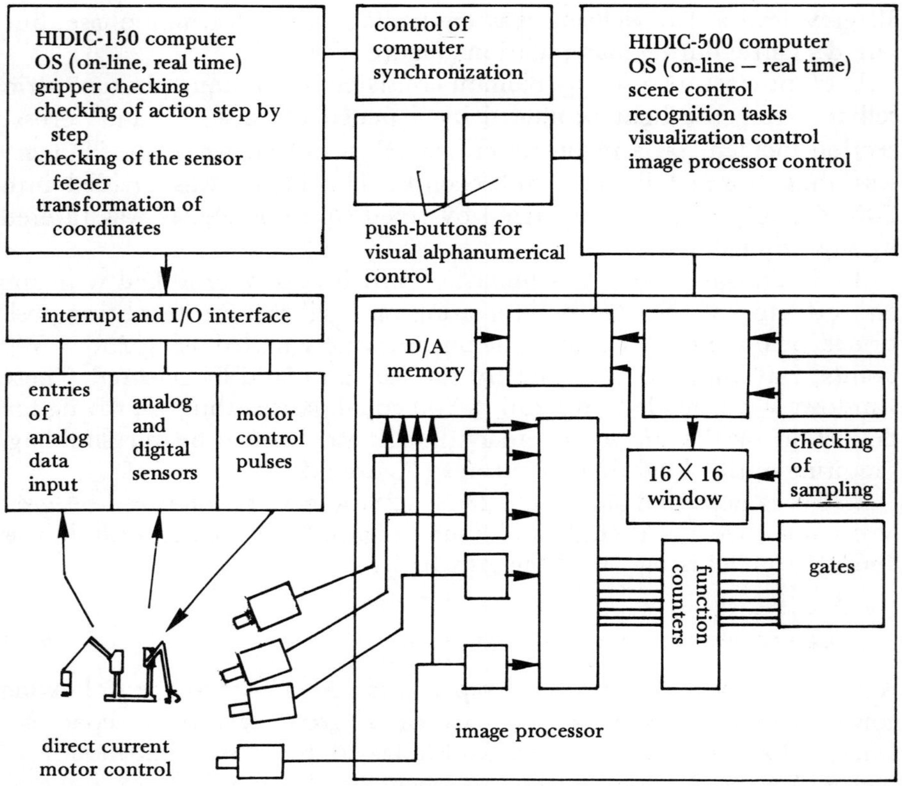

(a) The Kashioka system

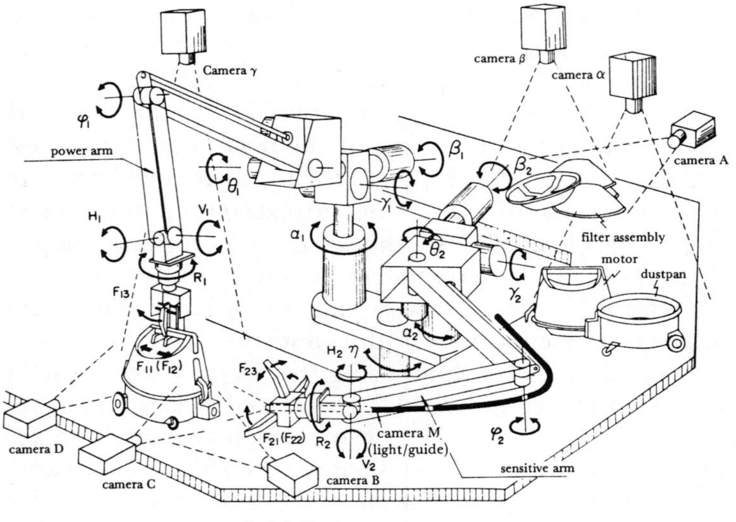

(b) Robot structure

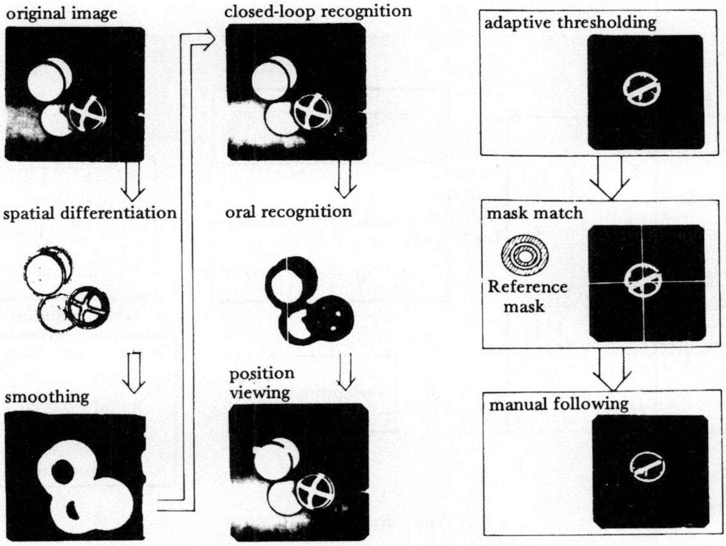

(c) Recognition of the filter and the engine casing

Figure 9.4. *The Kashioka experiment*

method of least squares and by assessing the error on all ten points. After distinguishing the objects from the scene, Yachida proceeded with a detailed analysis of every object with the help of an 11 X 11 window, and thus detected the outline. Therefore the objects were represented by their outlines in polar coordinates, their surface areas and the relationship of the surface area with the perimeter.

9.1.5 INTERPRETATION PHASE

In the experiment described by Agin[131] he identified the components by concentrating on four characteristic features: surface, maximum, minimum and average vector radius. The measurement-representation distance was estimated by the sum of the squares of the differences for these four features and the decision based on the thresholds connected with this distance. In case of conflict between various objects, other characteristic magnitudes were used (eg number of holes etc). He also used a binary decision tree, taking into account the group of characteristic features.

In the experiment a PDP 10 computer was used, working on shared time. Recognition time was two seconds. The other attempt made by Agin, using the array of photodiodes, was with a PDP 11/40 computer. With the same type of algorithm, recognition was made in 0.5 seconds.

For Hale the position of the drill was located using a mirror. The

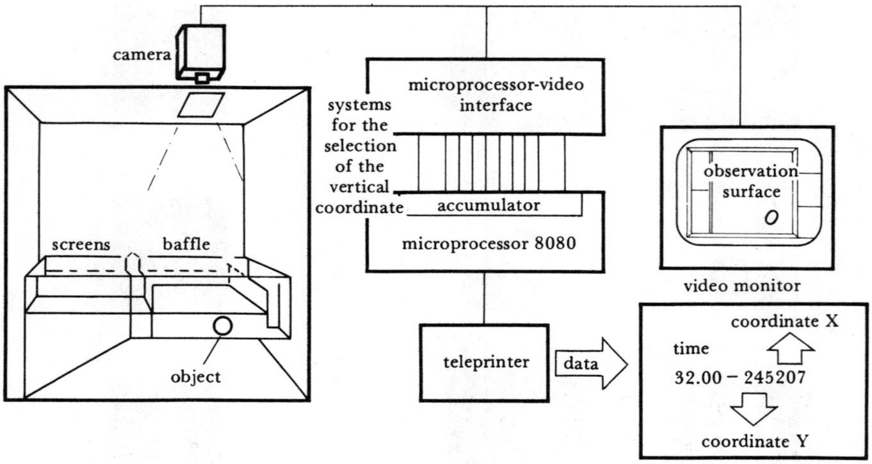

(a) System layout

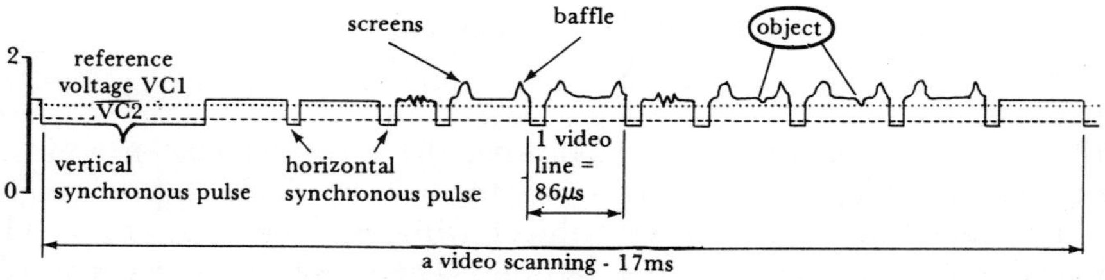

(b) Detection of the object

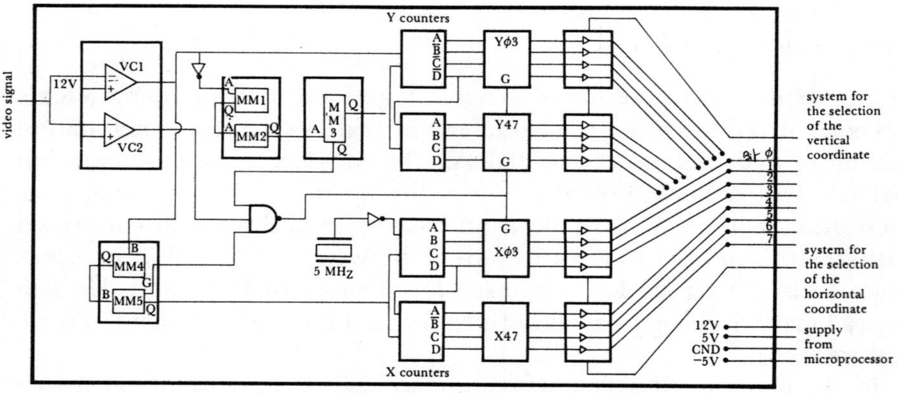

(c) Structure and control of counters

Figure 9.5. *The Lubinski experiment*

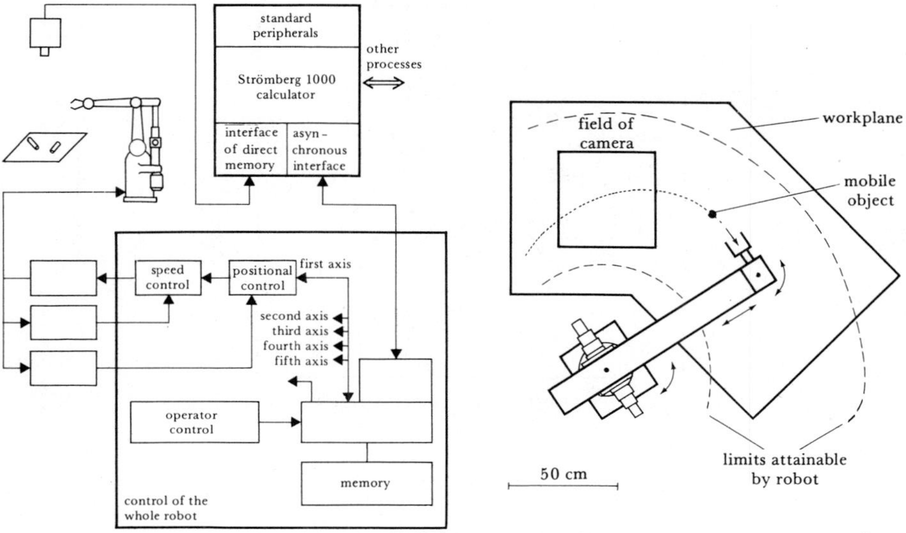

Figure 9.6. *The Niemi experiment*

algorithm identified an unpierced hole and controlled the drill. Recognition and positioning took six seconds, using a Honeywell 516 computer.

In Kashioka's work, parts were located by detecting the closed outlines of a predetermined shape (oval). The point of a vacuum cleaner part that was grasped was based upon the limits of this shape. In the assembly phase, the approximate position of the parts was known, because they were sited in the grippers of the two manipulators. The orientation was computed from guide-marks, easily detectable by the cameras. Two HIDIC 500 computers ran the system. One was charged with the acquisition and processing of visual information, as well as coordination of the action. The other handled the tactile information (monitoring the force employed in the assembly of parts) and the movements of the manipulators. The part recognition process took 7 to 12 seconds. Note that the general organization of the experiment rested on the parallel monitoring of recognition tasks and movement of the robots.

In the Lubinski experiment, an 8080 microprocessor monitored progress of the object's movements. In four seconds the coordinates of the object were registered. If, during this time, movement had been such that the object had moved through at least one-half of its dimension, the new coordinates were noted.

Niemi classified his cylinders according to their dimensions, using a modular program on a Stromber 1000 computer. A cylinder was grasped every 0.8 seconds.

The recognition method used by Perkins was more sophisticated. It

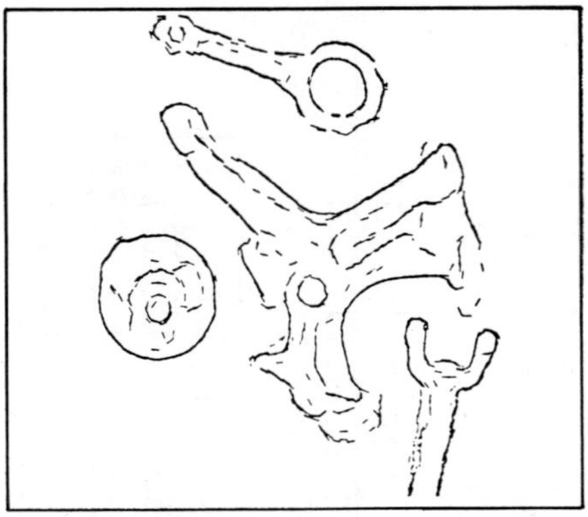

(a) Image after application of the Hueckel operation

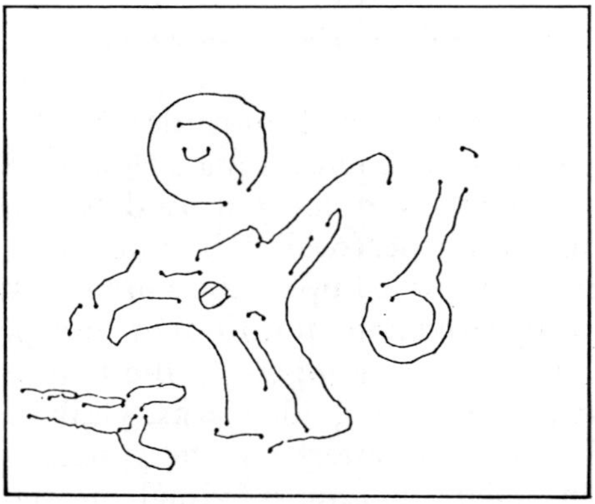

(b) Image after chaining

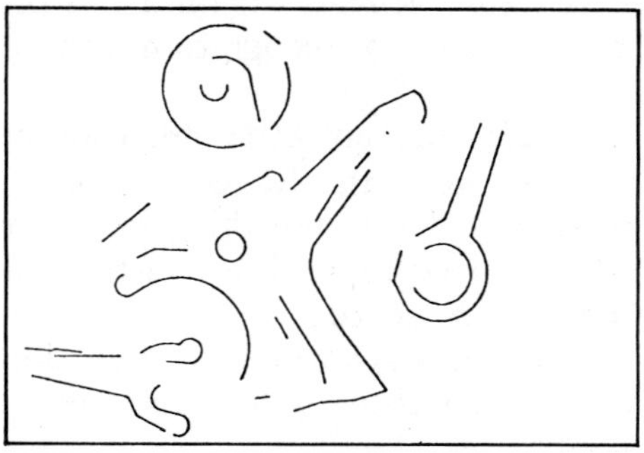

(c) Image after examination of straight-line segments and arcs of circles

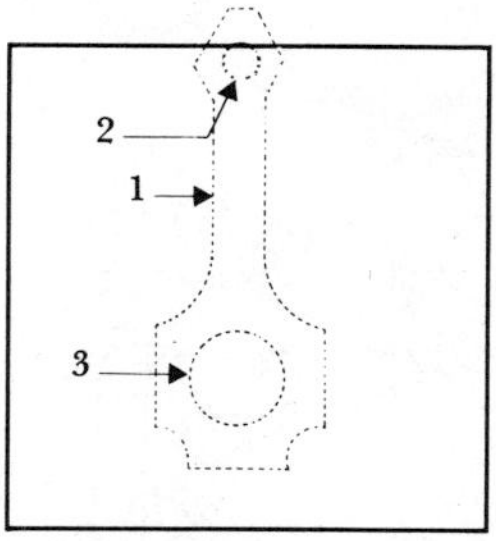

(d) Contours of model

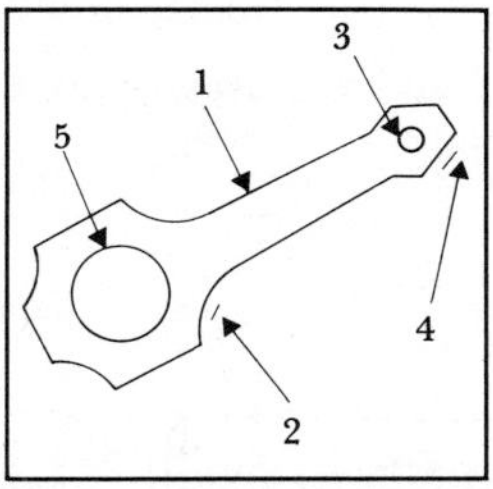

(e) Outlines of a chained model

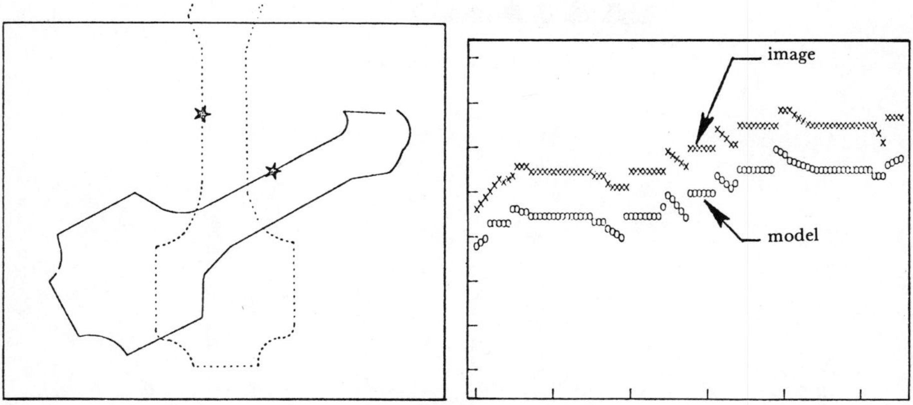

(f) Recognition

Figure 9.7. *The Perkins experiment*

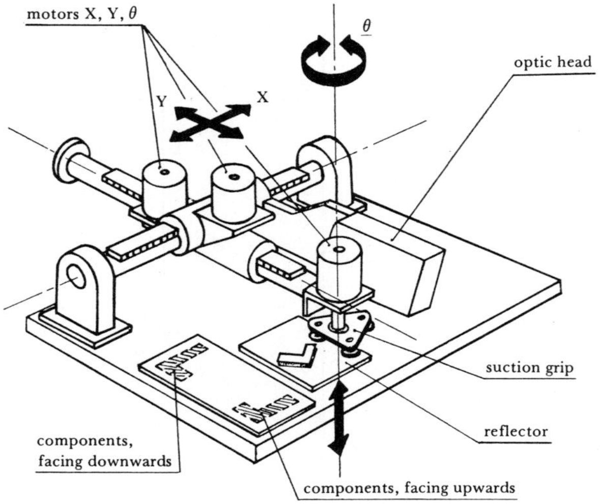

(a) System used by Salmon

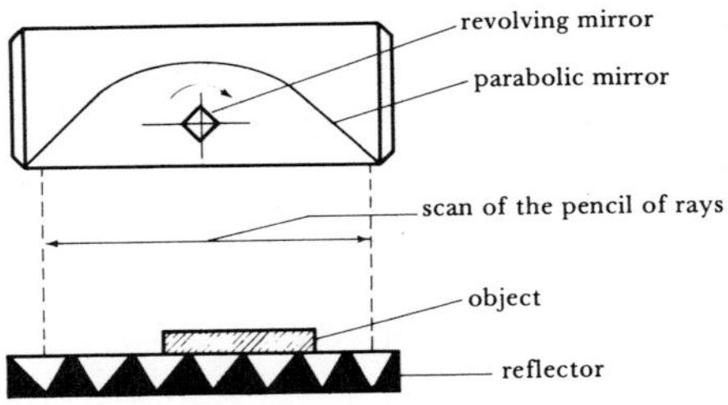

(b) The optic sensor

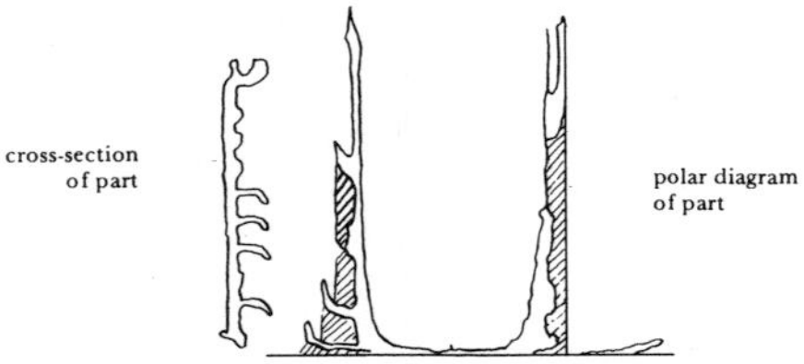

(c) Example of part (*left*): polar diagram of part (*right*).
Only hatched zones were used

Figure 9.8. *The Salmon experiment*

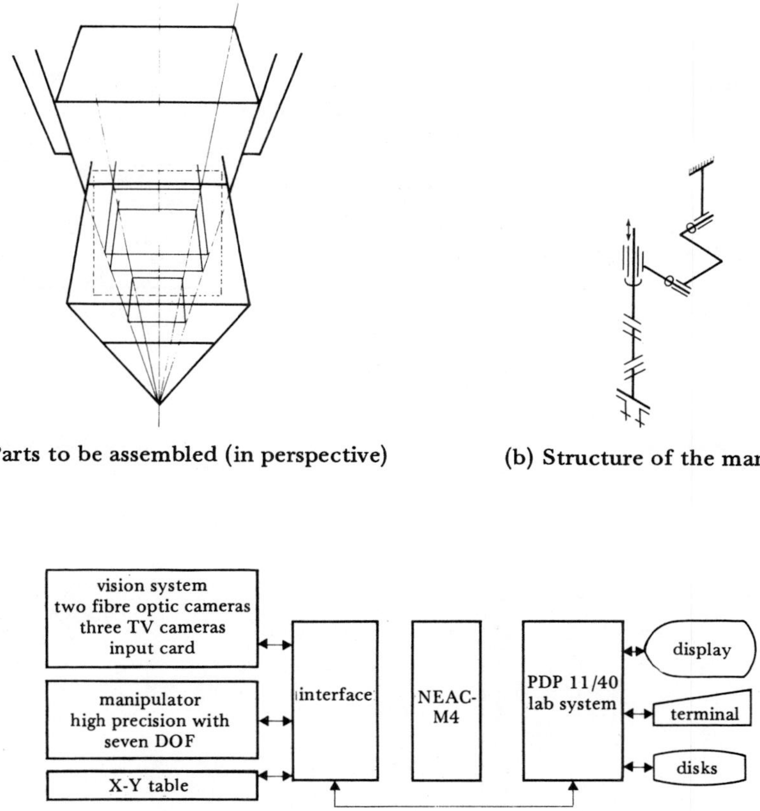

(a) Parts to be assembled (in perspective)

(b) Structure of the manipulator

(c) System layout

Figure 9.9. *The Tani experiment*

was carried out in three stages. First the outline of the detected object was compared with the recorded outlines, segment by segment. So the centre of gravity of the model was made to coincide with that of the image. Then, rotation of the model around this centre of gravity was carried out, until maximum coincidence was obtained. After that, resemblance between image and model beginning with the value of this maximum of coincidence was assessed. Approximately 20 seconds were needed using an IBM 370/168 for these three stages.

Like Perkins, Salmon also used template matching with a PDP 11/05 in which the arithmetic unit was wired. A recognition time of two seconds was needed; surface calculation needed 20 per cent, calculation of the centre of gravity 30 per cent, detection of the outline 10 per cent and use of the mask 40 per cent.

For Tani, the rectangular bar held by the robot was defined by its edges. The straight lines were made as parallel as possible, defining the

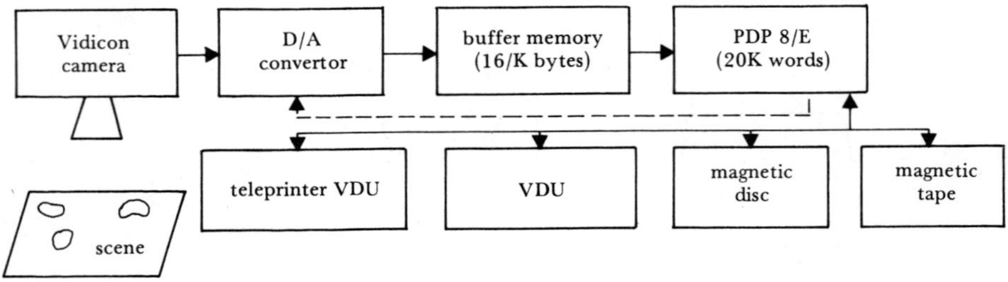

(a) General organization

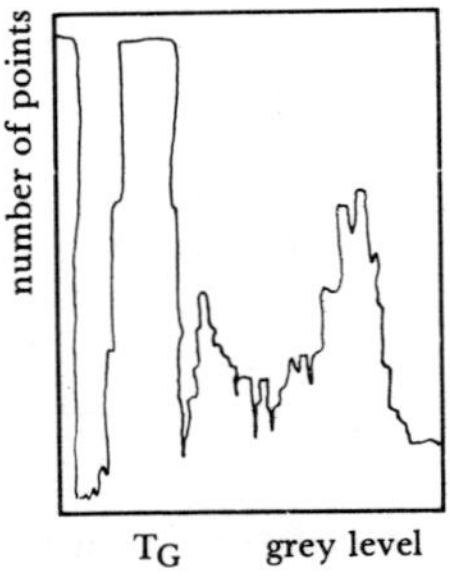

(b) Grey-level histogram

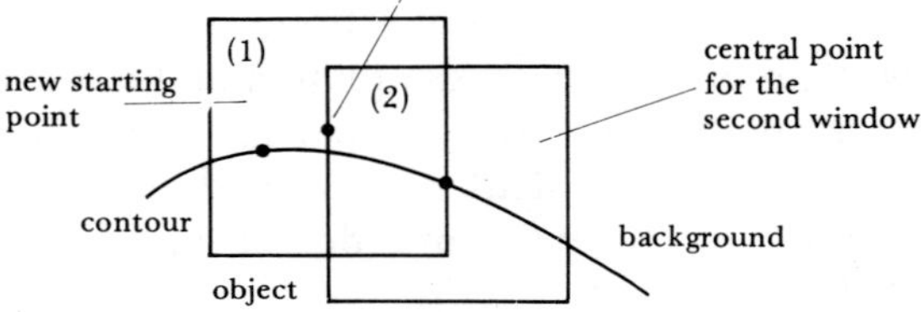

(c) Movement of the 11 × 11 windows for detection of the outlines

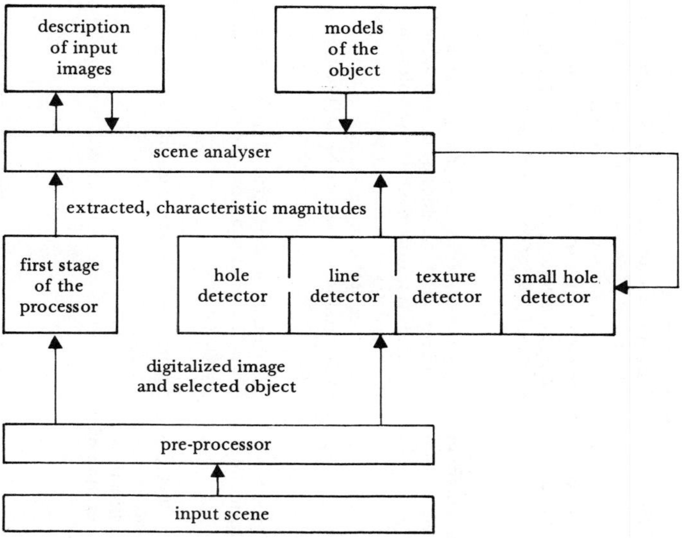

(d) Software structure

Figure 9.10. *The Yachida experiment*

edge of the hole and the block by making movements and measuring parallelism. The mechanical play between the bar and its bedding was approximately 0.5 mm and could not be less than 0.1 mm due to the camera's resolution. The processing of images was achieved using a PDP 11/40 and robot control using a NEAC–M4. Assembly took five minutes, of which three minutes were used in the vision process and its overheads. The processing of an image took eight seconds, but there was further processing required after each microdisplacement.

Yachida interpreted the image by comparing the polar description of the object with each of the stored models. If the various models proved convenient, other characteristics were extracted and compared eg 'waist' and position of holes, particular outlines. The computer used was a PDP–8/E, which took 20 to 90 seconds for recognition.

9.1.6 SYNTHESIS OF THESE EXAMPLES

In Tables 1 to 3, the characteristics of each experiment are stressed.[144] They summarize what has been discussed in previous chapters concerning robot vision:

(a) In all cases, the technique used is 2D (except the Kashioka technique). This means that the problem is really two-dimensional

Author	Objective	Acquisition phase	Representation phase
Agin[131]	Recognition, inspection of mechanical parts placed on conveyor belt. Flat parts, not touching	1. 120 × 120 TV camera, fixed, 32 levels of grey, binarization by fixed threshold; optimum lighting 2. Line of 128 photodiodes, one line every 1.2 mm; binarization by fixed threshold. Optimum lighting	Segmentation by zones of bits of same value. Extraction of characteristic parameters: surface, perimeter, circumscribed rectangle, relationship of perimeter with square root of the surface. Number of holes, position, surface. Vector radius from the centre of gravity to the perimeter: maximum, minimum and average
Agin[132]	Assembly of parts placed on a table. Fixed orientation	100 × 100 points CCD camera, mounted on the manipulator. Threshold of fixed binarization. Optimum lighting	As in the preceding case, except for the parameters: relationship of the sides of the circumscribed rectangle (measurement on the holes), deviation of the perimeter with respect to the circle, surface difference of the model-surface of the hole, distance of the candidate's centre to the position of the model
Hale and Saraga[134]	Piercing of holes on plate of printed circuit	TV camera 300 × 400 points, 32 grey levels. Binarization of fixed threshold. Optimum lighting	Sequential study of areas of 57 × 57 points. Use of a window of 11 × 11 for the examination of piercing zones
Kashioka *et al.*[135]	Assembly of parts placed on table. Partially or totally known position	Eight TV cameras: seven fixed, one linked to one of the manipulators by optic fibre: 256 × 256 points, 256 grey levels. Binarization by adaptive threshold. Augmentation of the contrast by spatial differentiation, filtering by smoothing	Segmentation in zones of bits of same value
Lubinski *et al.*[136]	Supervision of an object A single object; movement on a plane	TV camera, optimal lighting	Release of counters from horizontal and vertical position; detection of the video level for determination of the position of the object

Niemi *et al.*[137]	Study of recognition tracking and grasping tasks. Example with cylinders lying on a table	50×50 points CCD camera fixed on the work surface. Binarization by fixed threshold; optimum lighting	1. Recognition: segmentation by bits of same value; 3×3 window for detection of the outline. Determination of the C.G. and the orientation of the axis of the cylinders 2. Pursuit and grasping: segmentation, determination of the position of the object
Perkins[138]	Recognition of flat, mechanical parts, possibly partly hidden	Fixed TV camera, 256×256 points, 32 grey levels. Optimum lighting for taking the models but variable in the course of the experiment	Hueckel's operator for the detection of the outlines; constitution of chains of points; assignment of curve segments (straight lines or circle arcs) to each regular chain
Salmon[139]	Recognition of flat mechanical parts	Photocell and scanning system by light rays; 200×200 points, binarization by fixed threshold	Filtering by smoothing; detection of the outlines Representation by polar diagram
Tani *et al.*[140]	Assembly of prismatic parts	Three TV cameras: one fixed, two linked to the manipulator by optical fibres. Different degrees of resolution on a total of 4096 binary units per image. Binarization by fixed threshold, optimum lighting	Detection of the outlines, selection of the binary units belonging to a same segment of straight line Representation by a group of straight lines of the form: $y = ax + b$ or $x = ay + b$ with: $\|\partial\|S\|$
Yachida and Tsuji[142]	Recognition of flat mechanical parts or parts presenting a number of limited stable positions of balance	Fixed TV camera, 128×128 points in high resolution (a part of the image) or low resolution (the whole image); 64 grey levels, histogram of those for selection of binarization level	11×11 points windows for detection of the outlines. Calculation of the C.G. Polar representation and characteristic parameters: holes, particular outlines

Table 1. *Options available for the acquisition and representation phases*

Author	Preparation phase	Representation phase	Interpretation phase
Agin[131]	System-operator dialogue, indication of characters to be memorized. Checking of the development of the process by visualization bay. One image to be stored for each stable position of balance. Modular organization of the programs	Segmentation by zones of bits of same value; extraction of the characteristic parameters	Valuation of the object-model distance (sum of the squares of the differences of the measured parameters). Attempts with a binary decision tree, slightly improved results
Agin[132]	Not made clear	As before, but the measurements concern the holes in the parts to be assembled	Evaluation of the object-model distance
Hale and Saraga[134]	Not made clear	Sequential study of zones of 57 × 57 points, 11 × 11 window for the examination of holes. Determination of the position of the candidates	Evaluation of middle of segmented zones, determination of holes to be drilled
Kashioka *et al.*[135]	Not made clear	Segmentation of zones of bits of same value	1. Recognition: detection of specific forms (oval for the vacuum-cleaner's filter, circular with specific characters for motor block and for base of the engine) 2. Visual feedback: use of a mask to determine the distance of the parts in assembly with respect to the final position
Lubinski *et al.*[136]		Release of counters from horizontal and vertical position, detection of the video level for the position determination of the object (counter arrest)	Comparison of the actual position of the object with respect to its last position. Registration of this object if the movement is more than 50 per cent of the object's dimensions

Niemi *et al.*[137]	Not made clear	Segmentation: 3 × 3 window for detection of the outlines. Calculation of the C.G., orientation of the axes of the cylinders; and gripping: segmentation, determination of the object's position	Recognition: classification of the cylinders according to the middle. Tracking and grasping. Evaluation of the object's trajectory. Determination of a point of interception and orientation of the end effector with gripping in view
Perkins[138]	Creation models in optimal lighting conditions	Hueckel's operator for detection of outlines. Chains and coring of points; assignment of curve segments to each chain	Comparison with all the models segment by segment; choice of the nearest model, rotation of the model for maximum coincidence and determination of the orientation
Salmon[139]	Presentation of new parts under manipulation conditions. Control of recording of the model, not of interaction with the operator	Detection of outlines; representation by polar diagram	Comparison with all the models. Shifting as a function of the angles of the diagram up to a maximum coincidence for recognition and orientation
Tani *et al.*[140]		Detection of outlines; selection of the bits belonging to the same straight line segment	Examination of the hole with fixed camera, calculation of the block's orientation with respect to the hole
Yachida and Tsuji[142]	The operator can present new parts or modify the whole of description of those already memorized. The system registers the characters indicated and compares with the existing models. He eventually signals the indistinguishable models. The operator can therefore modify the description of the models	11 × 11 points window for detection of outlines, calculation of the C.G., polar representation and characteristic parameters: holes, specific outlines	Evaluation of the object-model distance by the polar representation. For similar objects, use of specific characteristics

Table 2. *Preparation and interpretation phases*

Author	Acquisition phase	Treatment time	Computer used
Agin[131]	Array of 128 photodiodes, 1 line for each 1.2 mm advance of the conveyor belt; binarization by fixed threshold, optimum lighting	1. Recognition: 500 ms on PDP 11/40, 2 s on PDP 10 2. Inspection: 300 ms on PDP 11/40, 2.4 s on PDP 10	Early PDP 10; PDP 11/40 in the most recent configuration
Agin[132]	100 × 100 points CCD camera, mounted on the manipulator; threshold of fixed binarization, optimum lighting	From 90 to 500 ms per image according to the complexity of the piece examined 5 s for an insertion (image analysis + calculation of the robot's movements)	PDP 11/40. Manipulator: UNIMATE 2000 B; 1.3 mm precision, six DOF
Hale and Saraga[134]	300 × 300 points CCD camera, 32 grey levels; binarization fixed threshold; optimum lighting	6 s for each drilled hole	Honeywell 516
Kashioka *et al.*[135]	Eight TV cameras, seven fixed, one linked to the manipulator by an optic fibre. 256 × 256 points, 256 grey levels; binarization by adaptive threshold. Augmentation of the contrast by spatial differentiation, filtering by smoothing	7 s for the recognition stages of the parts 150 ms in visual feedback	HIDIC 150 for checking of the manipulators. HIDI 500 for processing and image analysis
Lubinski *et al.*[136]	TV camera; optimum lighting		INTEL 8080
Niemi *et al.*[137]	50 × 50 points CCD camera, fixed above the workplane. Binarization by fixed threshold; optimum lighting	Recognition: 2 s for the first object, 800 ms for the next objects. Tracking: 1 to 2 s for evaluation of the trajectory	INTEL 8080 for checking of the manipulator; Stromberg 1000 for treatment and image analysis
Perkins[138]	Fixed TV camera, 256 × 256 points, 32 grey levels. Optimum lighting for the taking of models but variable in the course of the experiment	20 to 90 s for the recognition of an object	IBM 370/168

Salmon[139]	Photocell and scanning system by pencil of rays; 200 × 200 points, binarization by fixed threshold	500 ms for image taking; 1.8 s for treatment and image analysis	From equivalent power to the PDP 11/05 without unity of cabled, arithmetical calculations
Tani *et al.*[140]	Three TV cameras, one fixed, two linked to the manipulator by optic fibres. Different degrees of resolution according to the parameters of image taking; limited number of points to 4096. Binarization by fixed threshold, optimum lighting	8 s per image, three minutes for all the necessary images. Five minutes for an assembly	NEAC M4 for checking of the robot; PDP 11/40 for image processing
Yachida and Tsuji[142]	128 × 128 points, fixed TV camera for the whole image (low resolution), or for part of the image (high resolution). 64 grey levels, histogram for selection of binarization threshold. Optimum lighting	20 to 90 s per object	PDP 8/E

Table 3. *Time and means of calculation*

(eg the Hale technique) or it has been reduced to two dimensions by using prior information provided on the third dimension ie by employing one or more constraints.

(b) Consequently, the solutions involving a robotic vision system are essentially guided by the constraints that the designer has imposed or accepted for the particular problem eg:

– Constraints on the geometry of parts (Niemi's cylinders or Chien's cubes),
– Constraints on orientation (Hale's positioned boards, parts with the good side turned towards Agin's camera, vacuum cleaner motors with the good side turned towards the 'height' of Kashioka).
– Constraints on the position of parts.
– Constraints on the position of the sensors, always fixed, except in some of Agin's experiments.

(c) Consequently, not all experiments have equal 'flexibility' ie the same generality. Only the work of Perkins, Salmon, Agin and Yachida provide for new parts of unconstrained dimension to be stored without modification (or very minor modification) of the exploitation programs.

(d) Three basic methods, all deterministic, are used for recognition:

1. Exploitation of the object-model distance.
2. Mask matching technique adapted to each task.
3. Use of the decision tree.

They confirm the approach used in the preceding chapters.

(e) For sensors there are many television-type cameras and many solid-state cameras available.

(f) The degree of definition in the optic sensor stays low (from 2500 points for Niemi to 120,000 for Hale), which confirms the difference between 'classical' recognition and its interpretation in robotics.

(g) Calculation methods, recognition time and task duration are related, but the following should be noted: some researchers have used the equipment at their disposal without being able to modify it. Others have sought effective task execution without taking time and cost properly into consideration (this is true in the case of Perkins, using an IBM 370/168). And finally, some researchers sought to operate under industrial conditions (as in Kashioka's case).

So there is a range of aims and results. For example, with images digitized with an equal number of points, Kashioka carried out recognition in seven seconds using an HIDIC 150 and Perkins in 90 seconds with the IBM 370/168. Agin, working on the same parts as Perkins, only obtained definition corresponding to one-quarter of the points used for an image by the latter, and recognized the part in 0.5 seconds with a PDP 11/40. Not knowing the success rate, it is, of course, difficult to judge, but they are probably not very different.

From these examples, it is plain that a balance must be struck between the cost of representation techniques in keeping with those needed for industrial robots, having recognition time comparable with that of man, and a success rate compatible with industrial requirements.

(h) The Tani experiment was one of the most instructive. He wanted to know if assembly could be satisfactorily completed using only vision feedback from the environment. The results show that his method does not optimize the use of robots; it required five minutes with a PDP 11/40 and a NEAC–M4 to complete an assembly. Moreover, it depends on camera resolution, which requires precision to less than 0.1 mm. That totally confirms what has been said on the subject of assembly strategy: vision should serve as a rough guide to bring together the two parts and it is by stress feedback alone that the assembly may be best realized. Using this strategy, Tani would have been able to complete assembly in a few seconds.

(i) Scene lighting is a fundamental parameter, which affects calculation time, success rate and representation and recognition algorithms. This is why all the experiments quoted were carried out with planned lighting. Only Perkins, at the price of a complex representation, and Kashioka, with the help of an adaptive threshold, and Yachida using a histogram of the grey levels, could cope with limited lighting variations.

These examples cited above clearly reveal the problems and limits involved in the use of vision for robots.

9.2 Work involving depth of objects

Analysis of the literature on the subject of depth shows a confused situation. There have been few experiments concerning measurements of relief in robotics.

In fact, most emphasis has been on analysis of three-dimensional scenes (and designs of complex objects or three-dimensional scenes) – without much concern for calculation time and recognition rate or the competence of these analyses or designs when dealing with the problems presented by robots. So artificial intelligence techniques have to be considered.

Here attempts will be made to describe analysis of three-dimensional scenes of value for specific cases. The problem of three-dimensional localization by stereovision will be dealt with next. Finally, it will be seen how four cases have been carried into effect with artificial robots.

9.2.1 ANALYSIS OF THREE-DIMENSIONAL SCENES

This is a brief and very incomplete survey, which excludes methods relating to scenes viewed in the open or analysis based on colour, which is dealt with in the work of Kanadt.[145]

9.2.1.1 Analysis based on the outlines

Coding based on local constraints: This type of analysis assumes that the outlines have been extracted, so that the 'edges' of the scene's objects have been found. Therefore these edges are interpreted in a linguistic way by coding them and by coding their change of direction. This is the Huffman[146] technique of which an example is given in Figure 9.11.

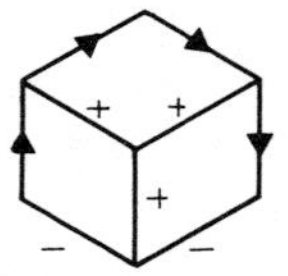

Figure 9.11. *Huffman's coding for a cube*

In these conditions, a node of the figure can only have 12 configurations (Figure 9.12).

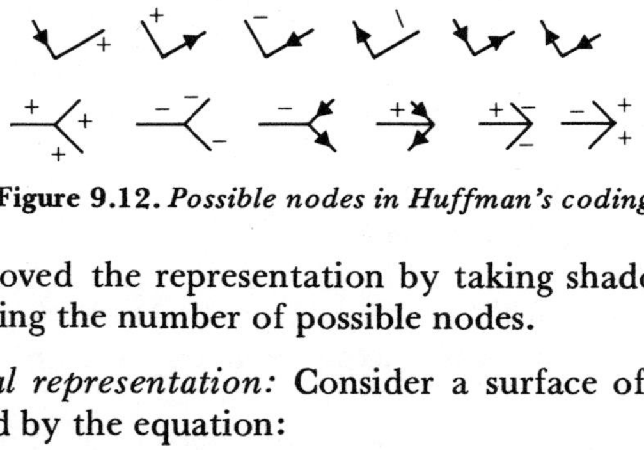

Figure 9.12. *Possible nodes in Huffman's coding*

Waltz[147] improved the representation by taking shadows into account and by increasing the number of possible nodes.

Coding of dual representation: Consider a surface of a scene (axes x, y, z), described by the equation:

$$z = ax + by + c \tag{9-1}$$

To this surface there corresponds in a three-dimensional space (u, v, w) a point (a, b, −c) in the original scene. The 2D image is obtained by projection of the dual space onto a surface. For example, the intersection of two planes

$$z = a_1 x + b_1 y + c_1 \tag{9-2}$$

$$z = a_2 x + b_2 y + c_2 \tag{9-3}$$

in the scene is represented in the dual space (u, v, w) by the straight line joining $(a_1, b_1 - c_1)$ and $(a_2, b_2 - c_2)$. When this line is projected onto the surface of the image (x, y) and onto the dual surface (u, v) of the surface of the image, two perpendicular straight lines are obtained if $x = u$ and $y = v$. Therefore the scene can be reconstructed using Huffman's method[148] if the image is codable.

Representation by systems of linear equations: The representation of an image consists of labelling areas, centres of these areas and writing the equations of the surfaces passing through n indexed centres. The intersections of these surfaces provides the construction of a polyhedral object.[149]

Curved objects: Following Turner,[150] one can use a code based on the nodes formed by the lines, by approximating the curve of these lines near the nodes. Turner suggests a code-book of 3000 possible junctions, thus generalizing Huffman's method.

Criticism of the approaches: At present, the above-mentioned methods are of little use in robotics. They:

1. assume the objects to be polyhedral, or they assume that the objects can be approximated by polyhedral outlines or surfaces described by simple equations. Although this is mathematically feasible for the simplified scenes considered by the authors, in practice the methods would lead to an extraordinary complexity of representation.
2. are based on the hypothesis that edges are always detected. This is not the case with current vision systems.
3. assume outlines to be perfect, but detection and smoothing of a real image has not been carried out.
4. seek to reconstitute the scene perfectly while only the key features are required by the robot.
5. do not impose a hierarchy on the various elements in the scene, a counter-intuitive step compared with human understanding of a scene.

Their range of application would be more appropriate, for example, to a graphic scene representation by a computer. They demonstrate the difficulty of reconstituting the third dimension, beginning with a unique, two-dimensional image. This is to be expected, because information on the third dimension is not easy to localize.

Tomography: Ultrasonic echography systems often lead to quite noisy images of various planes parallel to one another. It is useful to extract depth by detecting the outlines of the objects seen three dimensionally.

Rhodes *et al.*[51] have developed an algorithm which, in the neighbourhood of a pixel, considers the 26 nearest neighbours (three successive planes, nine points per plane, the 27 points forming a cube). The regions around 'seed' points are thus increased in number, by using acceptance tests for the new points considered, and the bulk of the object is reconstituted. This method imposes a constraint on the sensor itself which must be capable of providing images of successive planes, which TV or CCD cameras cannot do.

9.2.1.2 Analysis based on the measurement of size[152]

Quantitative measurements described in previous chapters have been based on measurement of distances by triangulation, transit time, stereovision, surface measurements and measurements of a completely different characteristic magnitude. These techniques have been referred to as K-2D and 3D.

9.2.2 RECONSTITUTION OF RELIEF BY STEREOVISION

There are two ways in which stereovision can be envisaged. One could be described as 'classic'. This is the measurement of a distance between a point of the scene and an indexed plane.

When P_1 is the image of point P on camera 1, and P_2 the image of P on camera 2 on one hand, and vector P_1P_2 and the directions of the radii P_1P and P_2P on the other hand, then the distance from P to camera 1 or to a plane whose relative position is known, can be calculated (Figure 9.13). Applying this principle to all points P, one can reconstitute the group of distances P_1P, P_2P etc and thus measure the depth.

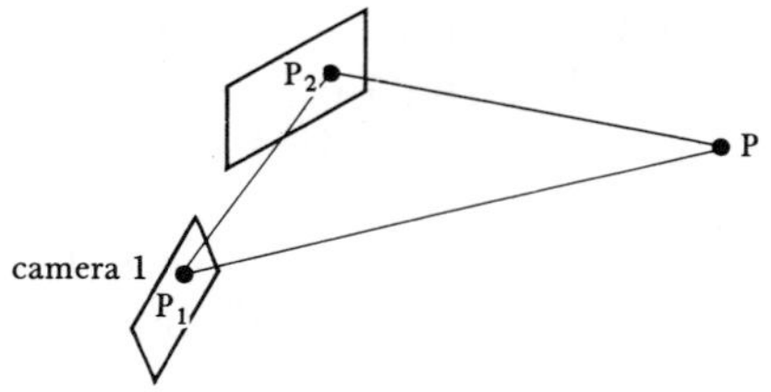

Figure 9.13. *Finding the position of P using two cameras*

Note that this works only if P can be easily seen by the two cameras, which implies only separating them slightly.

The great difficulty of this method lies in determining P, in the identification of P_1 and its corresponding point P_2 (and this for all possible points P). This is done in the calibration stage using known points P or having points P determined on one of the images by a human operator who examines areas of interest. Furthermore, it requires correlation between two areas thought to be connected, one to another, to detect two unique, corresponding points. If the images are noisy, this seriously complicates matters.

Gennery[153] and Yakimovsky[154] have developed systems intended for automatic vehicles ie for mobile robots moving in more or less known environments.

In manipulation processes, an object's orientation in space must be found. This is done by finding the reference axes related to the object's centre of gravity, with respect to a known framework.

Conventional stereovision is not much used. The other method, stereovision in fixed robots, is identical with the K-2D technique. Several images of the object are taken at different angles and not necessarily imposed beforehand. Exploitation of each image and the relationships between images (not necessarily expressed by a mathematical correlation point for point, as in classic stereovision) will provide reconstitution of the three-dimensional elements in the representation of the scene.

9.2.3 EXPERIMENTS USING THE K-TWO-DIMENSIONAL TECHNIQUE

9.2.3.1 The Driscoll and Levine experiment[155]

1. *Hardware:* In the Driscoll and Levine experiment automatic assembly involves the use of two arms: one for the assembly operation, and the other to move the optic sensor. Positioning of the sensor is carried out using an ultrasonic detector which places it opposite the object at a distance of between 25 and 450 mm. This sensor is, in fact, an infrared, active proximity gauge which scans the surface of the object.

2. *Coding images:* The objects form two-dimensional images limited by outlines. These outlines are classified as *non-singular* outlines ie without intersection, and *singular* outlines (Figure 9.14).

Figure 9.14. *The Driscoll classification of outlines*

Non-singular outlines are coded in terms of the area of the surface enclosed and the number and relationships between outlines, the latter by means of a graph (Figure 9.15).

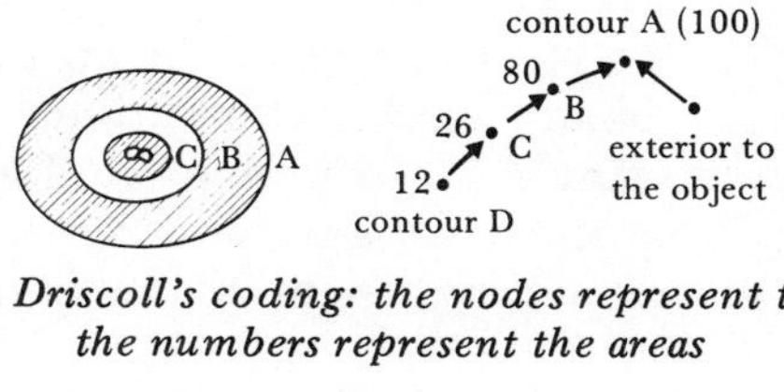

Figure 9.15. *Driscoll's coding: the nodes represent the outlines, the numbers represent the areas*

A coding has now to be established for all possible configurations seen from all possible viewpoints. This provides (imprecise) recognition of an object and its orientation. This is the case whatever the complexity of the object.

9.2.3.2 *The Kelley et al. experiment*[156]

In this experiment, a robot was required to handle several objects. The objects were assumed to be known and the problem reduced to separating them in the image and recognizing their orientation in space and their position.

The orientation was defined with respect to a reference mark linked to the fixed camera (Figure 9.16) by three angles of rotation about its axes, between 0 and 360°.

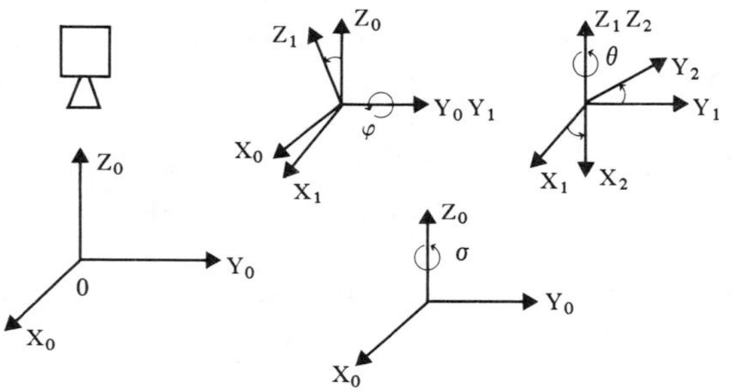

Figure 9.16. *The three rotations that define the orientation φ about Y_0 (pitch), θ about Z_1 (roll), σ about Z_0*

The matrix that expresses transfer of the coordinates from a point on the object in the reference axes linked to it, to its coordinates in the reference axes of the camera, is written:

$$\begin{pmatrix} X_0 \\ Y_0 \\ Z_0 \end{pmatrix} = \begin{pmatrix} C\varphi C\theta C\sigma - S\theta S\sigma & -C\varphi S\theta C\sigma - C\theta S\sigma & S\varphi C\sigma \\ C\varphi C\theta S\sigma + S\theta C\sigma & -C\varphi S\theta S\sigma + C\theta C\sigma & S\varphi S\sigma \\ -S\varphi C\theta & S\varphi S\theta & C\varphi \end{pmatrix} \begin{pmatrix} X_2 \\ Y_2 \\ Z_2 \end{pmatrix} \tag{9-4}$$

1. Characteristics extracted to describe the orientation of a component: A component is illustrated in Figure 9.17. The image is quantized, then thresholded to obtain a binary image. The outlines corresponding with the holes are extracted using a 2 × 2 mobile window and, by using the surface of the holes, the centre of gravity of the surface of the holes, and the maximum and minimum distances of this centre of gravity to the outline of the hole.

The parts being superimposed (Figure 9.18), the external outlines of the parts must be found. The line segments that allow the external outline to be found are formed by detecting significant changes in the

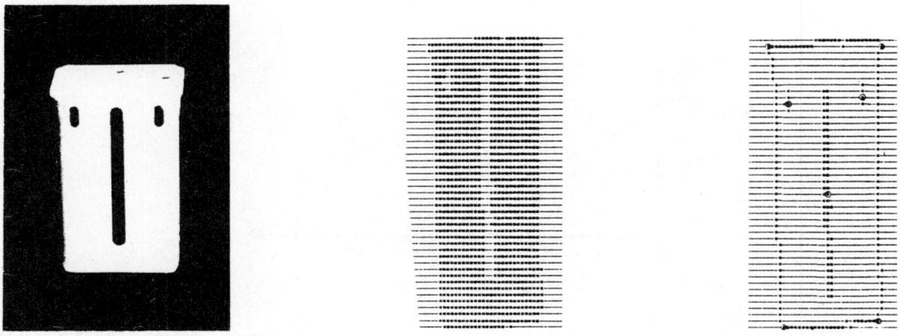

(a) (b) (c)

Figure 9.17. *(a) Analog image, (b) binary image, (c) contour and characteristic points*

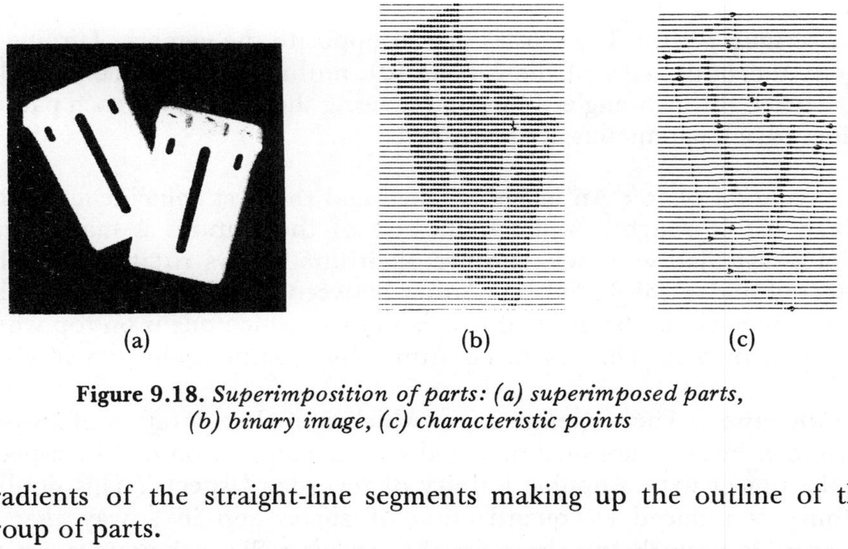

(a) (b) (c)

Figure 9.18. *Superimposition of parts: (a) superimposed parts, (b) binary image, (c) characteristic points*

gradients of the straight-line segments making up the outline of the group of parts.

If this curve is represented as a function of the gradient of the external outline, a number of peaks corresponding with sudden changes in direction of the outline (Figure 9.19) and a number of platforms corresponding with arcs of circles are found. Therefore the outline is presented as a succession of straight-line segments and arcs of circles (Figure 9.18c).

Parts are represented by:

(a) For each hole with a surface larger than threshold:
   1. The distance from the centre of gravity of the hole.
   2. The distance from the surface of the hole.
   3. Maximum and minimum vector radii.

(b) For the global outline of the part:
   1. The distance from the ends of each straight-line segment contained in the outline.

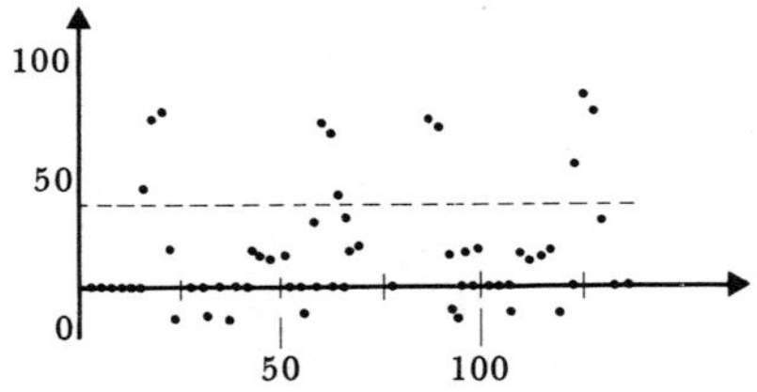

Figure 9.19. *Variation of the curve as a function of the arc length of the outline for the image of the object in Figure 9.17*

2. The distance from the end points and curve radii of each circle arc present in the outline.

2. Learning phase: The part is placed opposite the camera. Turning it about the optic axes of the camera (σ), nothing changes. Turning the part twice through angles θ and φ (eg using the robot), for each pair of values (θ, φ), parameters are registered.

3. Execution phase: An image is taken and the best coincidence in the stored list is sought. A hierarchic use of the features is made, thus gaining calculation time. Another algorithm allows rotation σ to be detected and a third, the translation between 0 and the centre of the part. The part can be grasped if it is known, which one is on top when two parts overlap. This can be confirmed by examining the grey levels.

4. Conclusion: This experiment, in which spatial orientation of known parts can be re-established, is based on the comparison of 'the aspect' of the object with a double infinity of recorded 'aspects'. This double infinity is reduced by quantization of angles and in a way that is compatible with the precision sought. This is a 2D technique in which the execution time and success rate might be improved by taking several images, as in the following experiment. Furthermore, it should be noted that, in the later work of Birk *et al.*[108] the bulk measurement problem has been resolved in quite a different way, namely by using a pneumatic sucker which picks up every part with a suitably flat surface, detected more simply using the same camera. The part grasped is placed in front of the robot in a horizontal plane, where the analysis required for a new seizure becomes simpler.

9.2.3.3 The Coiffet, Rives and Pot experiment[111,112,122]

In this experiment the objective was to recognize known objects, localize them and determine their orientation in space in order to manipulate them with a robot. The experiments were carried out on non-overlapping objects placed in a known horizontal plane. Everything was done not to hinder the extension of the experiments beyond these two constraints, but, nevertheless, these constraints were still observed. Figure 9.20 summarizes the approximations that governed the choices:

- Techniques independent of the nature of the objects.
- High success rate.
- Swiftness of execution.

When can the questions be answered?	Where are the objects?	Which are they?	Spatial orientation	Method for increasing the success rate	Method for increasing the speed
With one fixed camera	Always	Depends on the objects	Depends on the objects	Impossible	Impossible
With one mobile camera	Always	Always* with exception linked to the hidden side	Always	Several shots	Better strategy decreasing the number of shots
With a fixed camera and a mobile camera	Always	Always* with exception linked to the hidden side	Always	Several shots	Better distribution of the tasks between the two cameras parallelism

* In these specific cases, the object can be grasped and turned by the robot

Figure 9.20. *Approximations governing the choice of the system*

Experiment using mobile camera: The camera consisted of a CCD matrix comprising 64 × 64 points. The image was binarized by adaptive thresholding, as described in Figure 7.5. The outlines were extracted with the help of a Roberts' operator applied to the binarized image. The entire image and the outline of each object were memorized. The camera was fixed to the gripper of the robot (Figure 9.21).

1. Representation of an object: Dudani moments are used (section 7.3.3.2), calculated either for the whole image, or for the outline, if there are enough points:

$$V = (M_1, M_2; \ldots M_7) \tag{9-5}$$

These moments are characteristic of the image and vary according to the direction of the frame of the object. This direction can be characterized in a way similar to Birks' description using three angles indicating the relative orientation of the object with respect to the camera:

$$V(\varphi, \Psi, \xi) = [M_1(\varphi, \Psi, \xi), \ldots M_7(\varphi, \Psi, \xi)] \tag{9-6}$$

2. Learning and definition of φ, Ψ and ξ: Suppose the camera is

Figure 9.21. *CCD camera set on gripper of manipulator*

related to framework Casη and the object with guide-mark GIJK. These two guide-marks are represented in Figure 9.22 in their initial reference positions.

To describe an object, it is 'photographed' by turning successively about $\underline{K}(\xi)$, $\underline{J}(\varphi)$ and $\underline{I}(\Psi)$. Because the moments are invariant in a rotation of the image on itself, if the error due to the quantization of

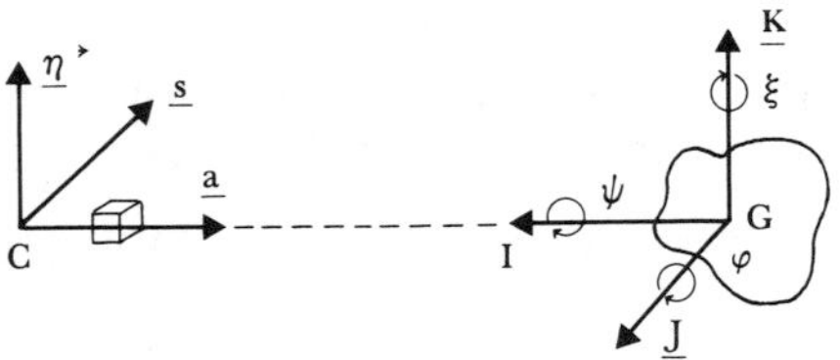

Figure 9.22. *Reference axes for the constitution of list F*

the image is neglected, a list F is registered which will represent a first group of characteristic vectors of the object:

$$V^1_{Ref} = [M_1(\xi, \varphi), \ldots M_7(\xi, \varphi)] \quad \textbf{(9-7)}$$

ξ and φ taking all possible values between 0 and 360°. The quantization chosen on φ and ξ is made by steps of 10°, which leads to 1296 vectors V^1_{Ref} present in list F for an object without symmetry.

A second series of vectors will allow characterization of Ψ by taking into account the positions of stable balance of the object. These are possible only when it is placed in a horizontal plane:

$$V^2_{Ref} = f(\xi, \varphi, \Psi) \quad \textbf{(9-8)}$$

ξ, φ, Ψ verifying the equations of stable balance for the object. These equations are, knowing ξ, of the form:

$$\begin{aligned} \sin \Psi &= k_1 \\ \cos \Psi \cdot \sin \varphi &= k_2 \\ \cos \Psi \cdot \sin \varphi &= k_3 \end{aligned}$$

k_1, k_2 and k_3 are constant. With the quantization chosen, there are 72 combinations per support side. If there are q support sides, there will be 72q vectors V^2_{Ref} that will form list G.

3. Automatic grasping of an object: Imagining, by way of example, an isolated, known object that is lying in the workplane in a known location, try to recognize its orientation in space.

The robot is going to look for the object, and place itself at such a distance that its image is complete on the sensor surface. The computer will evaluate vector V.

Before making a comparison with the V^1_{Ref} values, it must be remembered that reference-frame Casη and reference-frame GIJK have moved simultaneously. The camera sees the old reference-frame GIJL as a new reference-frame GI′J′K′ (known, because the camera's orientation with respect to the robot is known at all times). List G is put into the new reference-frame and list G_1 obtained. At that moment, the list of ξ values contained in G_1 gives the only possible orientations (due to stable balance conditions). From list F are only extracted a particular list of length 72q. This refers to V^1_{Ref} and is list F_1.

A comparison between the measured vector V and the group of vectors contained in F_1 can now be made in order to find ξ and φ.

There are several possible methods:

(a) Calculate a distance between the seven moments measured and those recorded, and decide that the minimum distance determines φ and ξ (a procedure identical to that of Birk). This is a procedure that may have to be abandoned because of:
- Image and quantization noise.

- Difficulty in finding the right weights of the moments to evaluate the measurement-reference moments.
- Calculation time, which increases with the order of the moments because of the increasing mathematical complexity of the formulae (section 7.3.3.2).

(b) The preferred solution is to:

- Reduce the dimension of vector V by keeping the first two components M_1, M_2 only.
- Reduce the number of possible vectors in F_1 (using a distance of a high threshold) without, at this stage, seeking the best candidate (reducing the few hundred vectors present in F_1 to a few dozen).
- Use the camera's mobility to look at the object from another angle and carry out cross-checking.

After this first image there are, therefore, a limited number of possible orientations (ξ, φ, Ψ) forming list G_2.

To reduce list G_2, a second image is taken, by turning the camera from a particular angle around the object. The same procedure is then carried out as for the first image, but the choice of orientations is limited to the list established in G_2, but evaluated ad hoc in the reference-frame G_3. In other words, a distance on the measured and registered M_1 and M_2 is used again, which reduces the possibilities in list G_3 to a short list forming list G_4 (some 10 to 20 remaining orientations).

At this stage one can choose between restarting the same camera manoeuvre or using a new criterion for reduction of G_4. The sign of variations on M_1 and M_2 between the first and the second image can be used. The measured signs ΔM_1 and ΔM_2 can be compared with the signs of the variations, ΔM_1^i ΔM_2^i, of the candidates' moments. This will again reduce the list of candidates to a few orientations forming list G_5.

At this level, a new image frame can be taken to extract the right candidate. It is simpler to use again a distance between ΔM_1 and ΔM_2 and ΔM_1^j ΔM_2^j calculated for the remaining orientations, and to select the orientation corresponding to the minimum distance.

Figures 9.23 and 9.24 show a schematic representation of the method using a test object: an irregular pyramid with a symmetry plane, and the flow diagram of an automatic seizure process.

4. Conclusion: The variations in M_1, M_2 etc are in the form of non-linear curves that have peaks (see Figure 9.25 where variations in M_1 are noted as a function of φ for values of ξ for the pyramid in Figure 9.23). If the right orientation is always chosen in list G_4, the use of variations in M_1 and M_2 in areas where they vary little – measurement errors being taken into account – can lead to discrimination errors in going from G_4 to G_5. Therefore either one or two moments (M_3, M_4) could be added so that there will certainly be differences in ΔM_i that

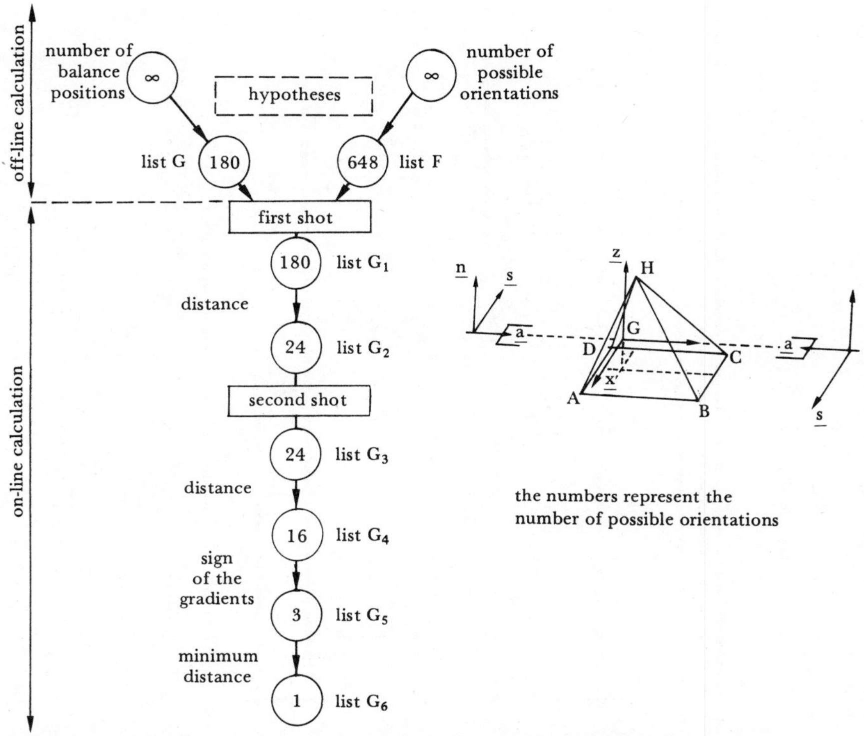

Figure 9.23. *Schematic representation of the method applied to the pyramid. The numbers represent the number of possible orientations*

are significant for at least two out of four moments. Or, the extent of the camera's movement between the first and second frame could be fixed to ensure that ΔM_1 and ΔM_2 have a definite significance. This could be done using a complex calculation, beginning with the registered variation model of M_1 and M_2, or by on-line adaptation (moving the camera with one value beforehand, measuring ΔM_1 and ΔM_2). If these divergences are insufficient, a new move is based on the values of ΔM_1, ΔM_2 etc.

Experiments using fixed and mobile cameras: Assuming the hypothesis of objects without overlap placed in a horizontal plane, it is known that a fixed camera positioned above the objects can give much information, but that it cannot resolve all problems without imposing constraints on the objects. It must be realized that such a fixed camera may be able to provide the correct solution in 95 per cent of cases, but the remaining five per cent are very important in actual optical application. Figure 9.26 shows an example of the incapacity of a fixed camera to distinguish between objects.

On the other hand, the mobile camera wastes time in the examination of the object, because it has to scan the scene in order to detect it.

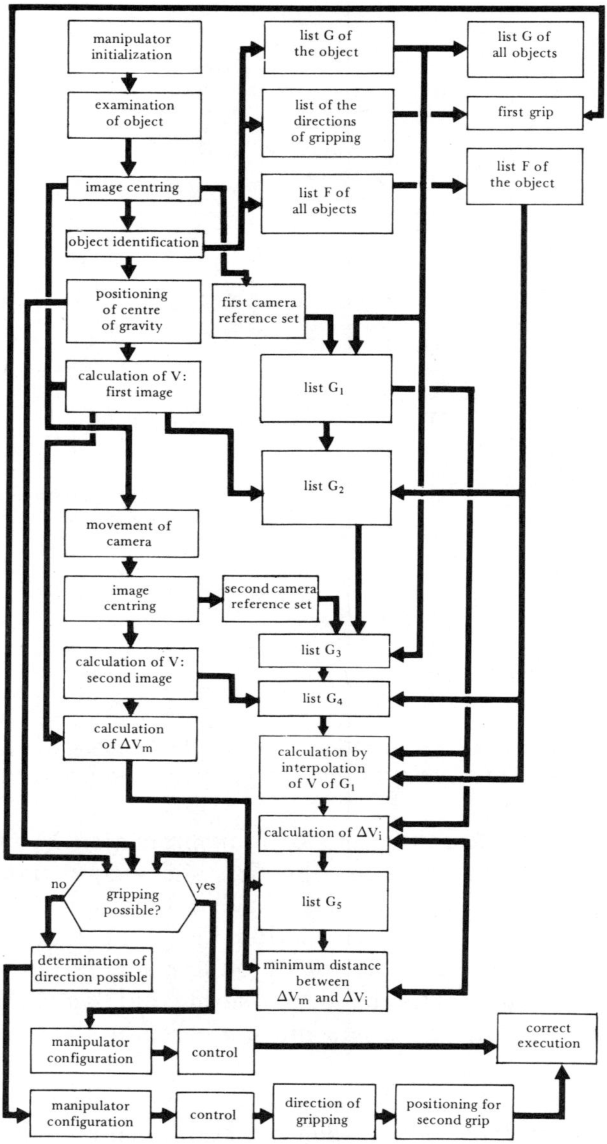

Figure 9.24. *Flow diagram of an automatic seizure process*

An apparently appropriate solution consists of using the fixed camera for recognition, placing and detection of orientation, and to remove all ambiguities that may arise with the help of a mobile camera.

1. Object recognition using a fixed camera: The sensor used is a Retican matrix of photodiodes of definition 100 × 100, which provides a binary image. The first phase consists of separating the objects contained in the scene using a connectedness algorithm which presents no problem, because the possibility of overlapping has been provisionally

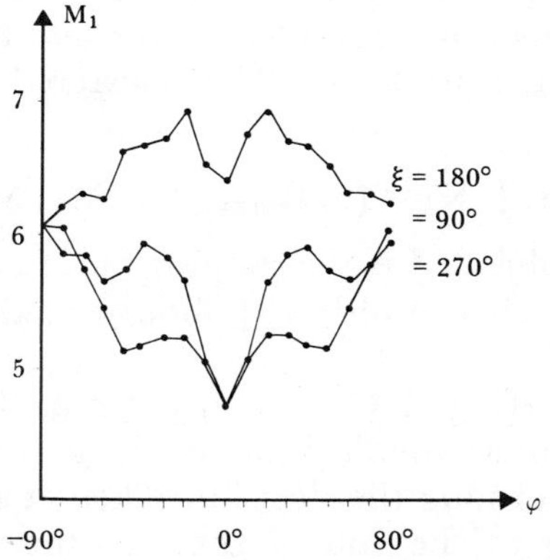

Figure 9.25. *Variation of M_1 for the pyramid in Figure 9.23*

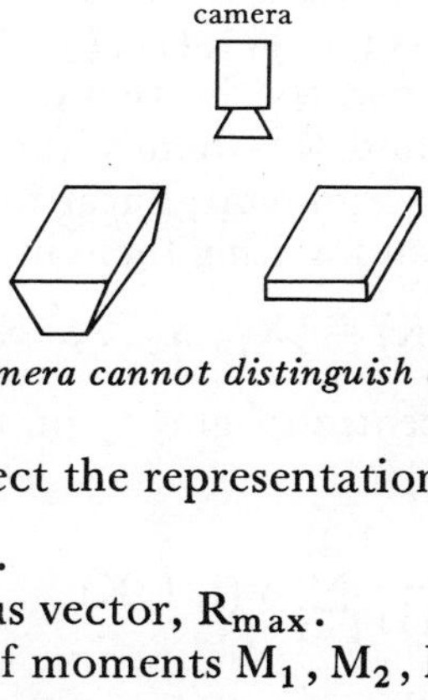

Figure 9.26. *The fixed camera cannot distinguish between these two objects*

eliminated. For each object the representation is based on:

– The area of image S.
– The maximum radius vector, R_{max}.
– The combinations of moments M_1, M_2, M_3.

The identification of the object is made using a discriminant tree, determined during the learning phase. The decisions, taken at every node of the tree, are defined by the sign of a linear combination of the parameters calculated for the object.

2. *Automatic formation of the decision tree:* In the learning phase, a list is created by presenting the different objects to the camera, for all possible orientations (in fact, one group per resting state), and by calculating the representation in each case. The number of objects the camera is shown, therefore, is not n, the true number of objects, but:

$$N = \sum_{i=1}^{n} q_i$$

q_1 being the number of rest states of real object no. i.

Consider an object placed on face J. If the object is turned by keeping it on its support side, there is only one possible rotation which would require an invariant stored representation. Because of spatial quantization of the image, this is not completely true and the values of S, R_{max}, M_1, M_2, M_3 are therefore noisy. The registered list is formed by the group of vectors:

$$V_{Ref}(I, J, K) = (S, R_{max}, M_1, M_2, M_3)_{I,J,K} \quad (9\text{-}10)$$

where I is the real object; J is the support face number for object I and K is the rotation number of object IJ about an axis normal to the plane on which it rests.

Now that list $V_{Ref}(I, J, K)$ has been created, the learning algorithm automatically determines the decision tree ie its structure and the values of the coefficients forming the decision taken at each node of the tree. From the beginning to the end of the tree there must be N parallel branches which will simultaneously provide identification of the object and support side.

The family of discriminant, linear functions is formed in such a way that for each pair of objects (I_1, J_1), (I_2, J_2) there is a function belonging to the family which separates the two objects. There are a number of methods available to find this family (gradient, factorial analysis, statistics etc).[86] Fisher's discriminant, linear functions are used.[157]

Changing the notations and writing them in more general form:

$$V(I, J, K) = [X_1, X_2, \ldots X_p]^T_{IJK} \quad (9\text{-}11)$$

For an object I, J the centre of gravity on the group of observations K is calculated using N(I, J):

$$g(I, J) = \frac{1}{N(IJ)} \sum_{k=1}^{N(IJ)} V(I, J, K) \quad (9\text{-}12)$$

The vector g(I, J) is defined:

$$g(I, J) = [g_1, g_2, \ldots g_p]_{IJ} \quad (9\text{-}13)$$

The centre of gravity of the group of observation is also calculated:

$$g = \frac{1}{\sum_{IJ} N(I, J)} \sum_{IJ} \left[\sum_{k=1}^{N(IJ)} V(I, J, K) \right] \quad (9\text{-}14)$$

$$g = [g_1^*, g_2^*, \ldots g_p^*] \quad (9\text{-}15)$$

The covariance matrix T of the group of observations (a square matrix of p × p dimensions) is defined by the generic element:

$$t_{r,m} = \sum_{I,J,K} \left[\frac{X_r(IJK) - g_r(IJ)] [X_m(IJK) - g_m(IJ)}{\sum_{IJ} N(IJ)} \right] \quad (9\text{-}16)$$

In these conditions, if two objects ($I_1\ J_1$) and ($I_2\ J_2$), that are in principle different *for the camera* are considered, Fisher's discriminant function is written:

$$[g(I_1\ J_1) - g(I_2\ J_2)]^T \cdot T^{-1} = F^{(I_2 J_2)}_{(I_1 J_1)} \tag{9-17}$$

Vector $F^{(I_2 J_2)}_{(I_1 J_1)}$ with p components defines an axis which maximizes the relationship of inter-class variance (inter-object) with intra-class variance (intra-object).

Now consider classes ($I_1\ J_1$) and ($I_2\ J_2$) and calculate the group of magnitudes for class ($I_1\ J_1$):

$$V_{1k} = [g(I_1\ J_1) - g(I_2\ J_2)]^T \cdot T^{-1}\ V(I_1\ J_1\ K) \tag{9-18}$$

that form a vector characterizing this class:

$$V_1 k = [V_{11}, V_{12}, \ldots V_{1,N(I_1 J_1)}]^T \tag{9-19}$$

Likewise for class $I_2\ J_2$:

$$V_2 k = [V_{21}, V_{22}, \ldots V_{2,N(I_2 J_2)}]^T \tag{9-20}$$

When the centres of gravity G_1 of V_{1k}, G_2 of V_{2k} are evaluated as well as the divergences of T_1 and T_2, one can classify according to one of two classes using the threshold defined by:

$$S = (T_2\ G_1 + T_1\ G_2)/(T_1 + T_2) \tag{9-21}$$

An observation V(IJK) (IJ being able to take values $I_1\ J_1$ and $I_2\ J_2$ only) will be classified as in $I_1\ J_1$ if:

$$[g(I_1\ J_1) - g(I_2\ J_2)]^T \cdot T^{-1} \cdot V(IJK) > S \tag{9-22}$$

It will be classified to be in $I_2\ J_2$ for the opposite case.

From the foregoing, the group of discriminant functions can be generated taking the classes 2 by 2: there are C_N^2 functions, if N is the total number of classes. This number can be considerably reduced by using knowledge that avoids useless comparisons. (Thus, for 23 objects, 34 functions only are needed instead of 253 functions predicted theoretically.)

This algorithm, implemented using microprocessor INTEL 8085, allows the decision tree to be worked out automatically, beginning with frames of the objects. So it is easy to go from one group of objects to another. For any four to six objects (eg Solex motor parts), the algorithm generates the tree in approximately two seconds and uses 14K words of 8 binary units.

3. Combined use of fixed and mobile cameras: The fixed camera can recognize various objects and their support side, except in specific cases of the type shown in Figure 9.26. Orientation of an object could also be used for recognition, but on the one hand the parameters contained in V_{Ref} (IJK) (equation 9-10) would have to be modified, and on the other hand, specific cases will remain in which orientation can only be known in the range of 180° (section 8.2).

Consequently, the mobile camera discussed in Chapter 8 is used to study spatial orientation. In these conditions, the fixed camera makes an image analysis by first indicating to the mobile camera where the selected object is. While the fixed camera identifies the object and its support side, the mobile camera is moved so as to take a photograph from which vector V of equation (9-5) is calculated. Then, the mobile camera knows with which object it has to deal and on which side it rests, so it does not have to choose from 72q combinations ξ, φ, Ψ but only from the 72 of them that become the dimension of list G_1. In these conditions, the procedure set out in section 9.2.3.3 gives sufficiently reliable results.

9.3 Choice of system

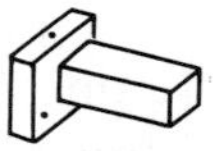

Figure 9.27. *Parts with recognition marks*

Consider an example in which parts shaped like the object shown in Figure 9.27 are distributed in quantity, the parts appearing to be identical. In reality, they differ by a few hundredths of a millimetre in their dimensions and are thus visually indistinguishable. Small distinguishing marks are to be found on each of the possible interfaces. The faces possessing the mark provide identification of the part. If the part is successfully grasped it must, when identified, be assembled with another one that carries the bore.

To execute the manipulation using a robot equipped with, for example, a two-fingered gripper and a vision system, the following procedure must be followed (not necessarily in the order shown):

1. Isolate one part from the others.
2. Be able to see it from all possible shooting angles.
3. Be able to take it with the robot's gripper.
4. Be able to grasp it in a direction needed for later assembly.

9.3.1 ISOLATION OF AN OBJECT

Eliminating the mechanical method – which can be used for sorting out parts by class and orienting them so that they can be grasped, the isolation of an object is based on image analysis. If the bulk is substantial, the difficulty of the problem is as has already been stated. If the shape is flat and if the images of the objects do not overlap, isolation is simple and easily carried out using a fixed camera positioned above the scene. This case will be taken as an example.

9.3.2 SEEING THE PART FROM ALL POSSIBLE DIRECTIONS

In this example it is necessary to be able to see the part from all directions in order to recognize it, because the marks could be anywhere and must be found. This is assuming either that the part moves, or that the camera moves. So there will be three main cases (Figure 9.28).

part on conveyor belt, fixed camera	part recognition	+	
	position	++++	
	orientation	++	
fixed part mobile camera	part recognition	++	
	position	++++	
	orientation	+++	
part grasped by robot for examination	first recognition attempt	fixed camera	++
		mobile camera	+++
	depositing of the grasped object for examination	second fixed camera	+++
		mobile camera	++++
	intermediary handling	second fixed camera	++++
		for examination	++++
	definitive handling	second fixed camera	++++
		mobile camera	++++

Figure 9.28. *Representation of the three methods of relative camera-object mobility and their chance of success in the most general problem, in which the objects are recognized by interpretable marks in any coded place of the object.*
++++ : theoretical certainty of success
+++ : success assured in numerous cases of specific objects
++ : theoretical success in specific cases
+ : very little chance of success

1. The part is carried on a *mobile support* eg a conveyor belt. The maximal rotations necessary for the recognition of the part are three rotations of 360° each ξ, φ and Ψ in Figure 9.22. If the conveyor belt presents the parts to the fixed camera, two limitations will present themselves:
   (a) Only the parts in a small segment $\Delta\xi$, $\Delta\varphi$ and $\Delta\Psi$ of the 360° of each rotation will be seen.
   (b) Even if the images do not overlap when the camera is above them, overlapping can occur as soon as the parts move away from the vertical and can necessitate an 'oblique' shot. The recognition rate will therefore be reduced considerably. On the

other hand, the position and approximate orientation of a chosen object will easily be established.

2. The parts are fixed and *the camera moves*. The angular limitations in the shooting directions will arise from:
   (a) The workplane in which the three rotations ξ, φ and Ψ are reduced from 360° to 180°.
   (b) The mobility of the camera which can mechanically rule out certain orientations of the camera with respect to the object (whether it is on the robot or on a nearby articulated system).
   (c) Overlapping of images from certain shooting angles. However, in this respect, the moving camera is nearly always better, providing larger values of $\Delta\xi$, $\Delta\varphi$ and $\Delta\Psi$ than with the conveyor belt and fixed camera case. So the recognition rate of the object and its orientation will be improved.
3. The parts are made mobile, with the robot's help. This indicates that a part has been grasped, which assumes two things:
   (a) That the position of the part to be grasped has been determined. This is quite easily achieved with a fixed camera.
   (b) That directions for which grasping is possible and a 'sign of graspability' have been determined.

9.3.3 'GRASPABILITY' OF AN OBJECT

The graspability of the object involves:

(a) The structure of the end effector.
(b) Its orientation with respect to the plane in which the object is at rest.
(c) The disposition of the other objects that might act as obstacles.

Example: Consider the object in Figure 9.29. With a two-fingered gripper of a width less than L, it cannot be seized when it is placed in configuration 4.

A gripping strategy must be developed, beginning with the use of visual information. A fixed camera will not always be able to supply that information; a mobile camera will be able to do it almost without fail. The relationships between gripping organ and object have been studied very little.[74]

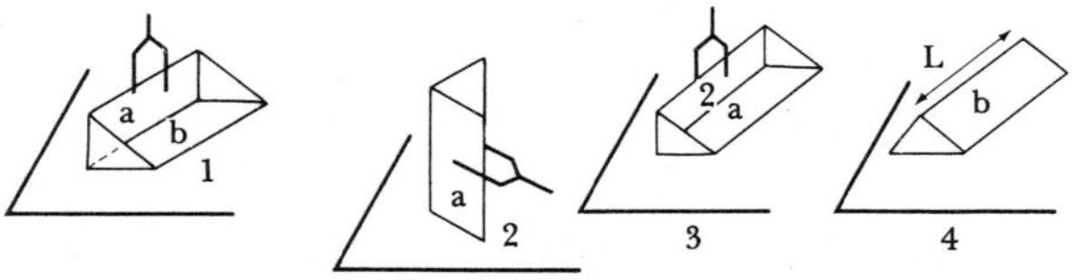

Figure 9.29. *Example of an object not graspable in position 4 with a two-fingered gripper of a width less than L*

Suppose the gripper succeeds in seizing the object. The robot turns the object before a fixed camera in all directions it is capable. This is to achieve recognition of the part. In practice, for image analysis, the gripper will hide parts of the object, which will generate the recognition process. The ideal is that the robot should place the object on a balance side in the workplane, in an area where it is alone. The fixed camera (not the same as the one which looked at the group of objects) will be able to examine it. There is, however, the risk that the recognition mark on the object cannot be seen on any of the balance sides. If it is examined by a mobile camera, the object will be seen from all directions and therefore recognition of the mark will be made without fail.

9.3.4 FINAL GRASPING OF THE OBJECT

The part has now been recognized and isolated. To carry out later assembly operations, it is necessary to grasp it correctly. If grasping is possible it must first be placed on the correct support side and then be taken to the assembly place. Note that, if the objects are not in a position in which they physically cannot be grasped, the last method (parts grasped by the robot) theoretically works in all cases, independently of the objects and their marks, if a mobile camera is used (K-2D technique). (A 3D feeler reduces the number of shots needed.) This method is therefore universal (and moreover resembles the human manipulation technique).

9.3.5 TEMPORARY PROBLEMS

The more general the method, the more it requires movements of robot or camera, and the more time-consuming it becomes. This is why, in real systems, specific properties of objects are best used and advanced information and constraints allowing the use of a fixed camera with mobile parts on a conveyor belt are preferred, introducing hidden time (Chapter 8). However, it may appear that the combined use of a fixed camera and a mobile camera enhances the generality of a vision system with a slight increase in cost.

Conclusion

In this second volume of the Robot Technology Series it was first of all necessary to focus the discussion. This meant concentrating on industrial applications (mainly of manipulation) or applications that could be industrialized quickly by using fixed robots ie in a confined environment and with few objects in it. It was thought proper to avoid arguments on terminologies and artificial intelligence. These will be presented in another volume. Finally, it was also decided to avoid inconclusive discussion of theoretical developments and concentrate on common sense, and basic ideas that should be a starting-point for research or applications in robot technology.

Mastering environment-robot interaction is the key to robot technology. A robot capable of adapting to a task, without human intervention, would change the very *nature* of existing industrial robots and, therefore, the nature of production methods, if not of manufactured products.

We have seen that the problems found in environment-robot interaction have some way to go before being fully resolved. A contributory factor is the lack of precision surrounding the terms 'environment' and 'task'. In fact, the research worker is tempted to regard anything from lunar soil to a heap of potatoes as environment and to initiate very general projects, so that they bear little relationship to the real environment and the existing robot. Although theory may be appealing, this book has appealed to the view that with the example of classical pattern recognition, blind application to every case would necessitate an almost total revision of theory to take technological and economic factors (duration, success rates, costs etc) into account. Hence realistic constraints on the environment have been discussed; and despite that, not everything has been achieved.

Another element in mastering interaction lies in showing to advantage the synergy between the robot's various components, namely the different articulations, the different internal and external sensors and possibly the mobilities of the tool manoeuvred by the robot. The term 'synergy' implies coordination and use of mechanical or informational redundancies to ensure an optimal execution of the task with respect to often highly contradictory criteria. The literature has often stressed the eye-hand coordination, but perhaps an eye-hand-touch coordination should be envisaged.

References

1. Farreny, H. *A System for the Expression and Resolution of Problems for Robot Control* Thesis, Toulouse, 1980.
2. Kaiser, C. *Physiology* Volumes 1 and 2, Editions Médicales Flammarion, 1969.
3. Thurston, H. *The Physical Phenomena of Mysticism* Gallimard, 1961.
4. Liegeois, A.; Khalil, W. The dynamics of a class of electrically actuated and cable-driven manipulators. *Proceedings of IUTAM Symposium* Munich, 1977.
5. Bruhat, G.; Foch, A. *Mechanics* Masson, 1961, 6th edn.
6. Watson, P.C.; Drake, S.H. Pedestal and wrist force sensors for automatic assembly. *Proceedings of 5th International Symposium on Industrial Robots* IIT Research Institute, Chicago, September, 1975, pp. 501–511.
7. Binford, T.D. Sensor system for manipulation. *Proceedings of 1st Conference on Remotely Manned Systems* ed. Heer, E., 1973, pp. 283–291.
8. Goto, T.; Inoyama, T.; Takeyasu, K. Precise insert operation by tactile controlled robots: HI-T-HAND, Expert 2. *Proceedings of 4th International Symposium on Industrial Robots* Tokyo, 1974, pp. 209–218.
9. Rosen, C.A.; Nitzan, D. Development in programmable automation. *Manufacturing Engineering* September, 1975, pp. 26–30.
10. Groome, R.C. *Force Feedback Steering of a Teleoperator System* Draper Laboratory, Cambridge, Report no. T. 575, 1972.
11. Hill, J.W.; Sword, A.J. Manipulation based on sensor directed control: an integrated end effector and touch sensing system. *Proceedings of 17th Annual Human Factor Society Convention* Washington DC, 1973.
12. Bejczy, K. Smart sensors for smart hands. Remote sensing of Earth from space: role of smart sensors. *Progress in Astronautics and Aeronautics* 1980, **67**.
13. Whitney, D.E.; Nevins, J.L. The remote center compliance – what can it do? *Robotics Today* Summer, 1979.
14. Inoue, H. Computer controlled bilateral manipulator. *Bulletin of Japanese Society of Mechanical Engineering* 1971, **14**, 199–207.
15. Khalil, W. *Contribution to Automatic Manipulator Control Using a Mathematical Model of the Mechanisms* Thesis, Montpellier, 1978.
16. Borrel, P. *Model of Manipulator Response. Application to Their Performance and Their Automatic Control* Thesis, Montpellier, 1979.
17. Bolles, R.C.; Paul, R. *The Use of Sensory Feedback in a Programmable Assembly* Memo 220, Stanford Artificial Intelligence Project, Stanford University, 1973.
18. Giraud, A. Private communication, 1981.
19. Coiffet, P.; Chirouze, M.; Vacant, A.; Flaissier, M. An experimental approach for a force controlled robot in an industrial shoe factory. *Proceedings of 10th Conference on Industrial Robots* Milan, 1980.
20. Nevins, J.L.; Whitney, D.E. Research on advanced assembly automation. *Computer* December, 1977, pp. 24–38.

21. Watson, P.C. A multidimensional system analysis of the assembly process as performed by a manipulator. *Proceedings of 1st North American Robot Conference* October, 1976.
22. Stute, G.; Erne, H. The control design of an industrial robot with advanced tactile sensitivity. *Proceedings of 9th International Symposium on Industrial Robots* 1979, pp. 519–527.
23. Cassinis, R. Sensing systems in supersigma robots. *Proceedings of 9th International Symposium on Industrial Robots* Washington, DC, 1979, pp. 437–448.
24. Peruchon, E. *Contribution to Carrying Out Artificial Tactile Functions Based on Analysis of the Human System* Thesis, Montpellier, 1979.
25. Bannister, L.H. Sensory terminals of the peripheral nerves. In *The Peripheral Nerve* ed. London, D.N., Chapman and Hall, London, 1976.
26. Alnaes, E. Static and dynamic properties of Golgi tendon organs in the anterior tibial and solens muscles of the cat. *Acta Physiologica Scandanavia* 1967, **70**, 176–187.
27. Munger, B.L. The intraepidermal innervation of the snout skin of the opposum. A light and electron microscopy study, with observations on the nature of Merckel's. *Journal of Cell Biology* 1976, **26**, 79–96.
28. Hermann, H.; Cier, J.F. *The Central Nervous System* Masson, Paris, 1968.
29. Umetani, Y. Principles of piezoelectric micromanipulator with tactile sensitivity. *Proceedings of 8th International Symposium on Industrial Robots* Stuttgart, 1978.
30. Kinoshaita, I.; Aida, S.; Mori, M. A pattern classification by dynamic tactile sense information processing. In *Pattern Recognition* Volume 7, Pergamon Press, Oxford, 1975.
31. Page, S.; Pugh, A.; Heginbotham, W.B. Novel techniques for tactile sensing in a three-dimensional environment. *Proceedings of 3rd Conference on Industrial Robot Technology* University of Nottingham, 1976.
32. Goldgewicht, M. A shape identification system. *Machine Moderne* June, 1974.
33. Takeda, S. Study of artificial tactile sensors for shape recognition algorithms for tactile data input. *Proceedings of 4th International Symposium on Industrial Robots* Tokyo, 1974.
34. Clot, J.; Falipou, J.P. *Tactile sensors* ANVAR brief 75.34.225.5, November, 1975.
35. Briot, M. *Robot Stereognosis. Application in Solid Sorting* Thesis, Toulouse, 1977.
36. Stojilkovic, S.; Saletic, R. Tactile pattern recognition by Belgrade hand prosthesis. *Proceedings of 5th International Symposium on External Control of Human Extremities* Dubrovnik, 1975.
37. Clot, J.; Stojilkovic, S. Integrated behaviour of artificial skin. *IEEE Transactions on Biomedical Engineering* July, 1977.
38. Okada, T.; Tsuchiya, S. *Object Recognition by Grasping* ETL-AIR Group Technical Note 18, 1976.
39. Briot, M.; Renaud, M.; Stojilkovic, S. An approach to spatial pattern recognition of solid objects. *IEEE Transactions on Systems, Man and Cybernetics* 1978.
40. Inoue, H. *Tactile Pattern Recognition for Manipulating Objects* ETL-AIR Group Technical Note 2, 1972.
41. Espiau, B. *The Use of Proximity Switches in Teleoperation* IRISA Report, October, 1980.
42. Bauzil, G.; Vaisset, M. The robot ultrasonic navigation module. *Deuxième Journée de Robotique* Rennes, May, 1981.
43. Johnston, A.R. *Optical Proximity Sensors for Manipulators* NASA Technical Memorandum 33-612, 1973.

44. Anon. *Optical Ranging Systems, Infrared Position Sensor* JPL Internal Report, 1976.
45. Ruget, G.; Espiau, B. *Study and Use of Proximity Detectors Using Fibre Optics* Spartacus Project, Final Report of IRIA, Contract 76-77-131, October, 1978.
46. Van Trees, H.L. *Detection, Estimation and Modulation Theory* Wiley, New York, 1968.
47. Gibson, J.J. *The Senses Considered as Perceptual Systems* Houghton Miffen Co., Boston, 1966.
48. Gregory, R.L. Mechanisms of perception. In *Computer Vision and Sensor-Based Robots* Plenum Press, New York, 1979.
49. Helmholtz, H.L. *Handbook of Physiological Optics* JPC Southall, Dover, 1962.
50. Aleksander, I. *et al.* Computer vision for industry: WISARD and the like. *Digital Systems for Industrial Automation* 1982, 2(1).
51. Pimonow, L. *Introduction to Bioinformation* Volume 1, Practical School of High Level Studies, Paris, 1975.
52. Reddy, D.R.; Hon, R.W. Computer architectures for vision. In *Computer Vision and Sensor-Based Robots* Plenum Press, New York, 1979.
53. Kato, H. A new integrated robot eye for colour discrimination. *Proceedings of 7th International Symposium on Industrial Robots* Tokyo, 1977, pp. 127–134.
54. Germain, F. Problems of loose objects in three dimensional perception. *Deuxième Journée de Robotique* Rennes, May, 1981.
55. Borelly, J.J.; Laurgeau, C. An autonomous sensor in tridimensional imagery. *Deuxième Journée de Robotique* Rennes, May, 1981.
56. Tienwen Chien, R.; Snyder, W.E. Hardware for visual image processing. *IEEE Transactions on Circuits and Systems* June, 1979.
57. Horn, B. *The Image Dissector 'Eyes'* (Project MAC, MIT, Vision Flash 16) Cambridge, March, 1971.
58. Dreyfus, M.G. Visual robots. *Industrial Robot* December, 1974.
59. Gille, B. History of technology. In *Encyclopédie de la Pléiade* 1978.
60. Hawkins, J.K.; Munsey, C.J. Image processing by electron optical techniques. *Journal of Optical Society of America* 1967, 57, 914–918.
61. Vidal, J.J.; Koludjian, G.H. Evaluation of flying spot scanners as image processing devices. *Proceedings of the Society of Photo Optical Instrumentation Engineers Seminar* South Fallsburg, New York, 1969.
62. Butler, J.W. Alice: a general purpose image processing system. *Proceedings of the Informal Colloquium on Polly Devices* Argonne National Laboratory, January, 1972.
63. Poppelbaum, W.J. *Computer Hardware Theory* Macmillan, New York, 1972.
64. Singer, B.; Kostelec, J. *Theory, Design and Performance of Low-blooming Silicon Diode Array Imaging Targets* Philips Laboratory, Briarcliff Manor, New York, 1972.
65. Palmieri, G. Image devices for pattern recognition. *Pattern Recognition* 1972, 3.
66. Laporte, M. Load transfer systems and their use in Earth observation. *Proceedings of International Conference on Earth Observation from Space and Management of Planetary Resources* Toulouse, March, 1978.
67. Belan, J.P.; Louineau, A. Charge transfer image analysers – potential application in television. *Radiodiffusion-Television* April–May, 1977.
68. Rives, P. *Contribution to the Development of Shape Recognition in Robotics* Thesis, Montpellier, 1981.
69. Casasent, D. Spatial light modulators. *Optical Engineering* 1978, 17, 307–384.

70. Gara, A.D. Optical computing for image processing. In *Computer Vision and Sensor Based Robots* Plenum Press, New York, 1979.
71. Jacobson, A. Real-time optical data processing device. *Information Display* 1975, 12.
72. Thompson, B.J. Hybrid processing systems – an assessment. *Proceedings of IEEE* 1977, 65.
73. Shirai, Y. A step towards context sensitive recognition of irregular objects. *Computer Graphics and Image Processing* 1973, 2.
74. Albus, J.S. Measurement and control for adaptive robots. *Proceedings of 10th International Symposium on Industrial Robots* Milan, 1980.
75. Johnston, A.R. *Infrared Laser Rangefinder* NASA New Technology Report NPO-13360, JPL, Pasadena, California, August, 1973.
76. Nitzan, D.; Brain, A.; Duda, R. The measurement and use of registered reflectance and range data in scene analysis. *Proceedings of IEEE* 1977, 65.
77. Rosen, C,. Nitzan, D. *Computer* December, 1977, pp. 12–24.
78. Armstrong, W.W.; Cadieux, P. The SIV development system and its application in scene analysis. *Proceedings of the ENSM Conference* Besançon, 2 July, 1981.
79. Hannah, M.J. *Computer Matching of Areas in Stereo Images* Memo AIM-239, Stanford Artificial Intelligence Project, Stanford University, July, 1974.
80. Haine, F. *Study and Production of Ultrasonic Imagers With Electronic Sectorial Sweepers* Thesis, Valenciennes, November, 1980.
81. Attal, J. Acoustic microscopy: imaging microelectronic circuits with liquid metals. *Scanned Image Microscopy* Academic Press, London, 1980.
82. Alais, P. 2D imaging versus holography and 3D imaging. In *Electrocardiology* ed. Bom, N., 1977.
83. Earnest, L.D. *Choosing and Eye for a Computer* Memo 51, Stanford Artificial Intelligence Project, Stanford University, 1973.
84. Anon. *Advances in Plumbicon Tube Design* Philips Laboratories, Briarcliff Manor, New York, 1971.
85. Simon, J.C.; Guiho, G. Picture recognition and transformations by computer. *Pattern Recognition* 1971, 3, 169–178.
86. Duda, R.O.; Hart, P.E. *Pattern Classification and Scene Analysis* John Wiley & Son, 1973.
87. Wigglesworth, V.B. The life of insects. *Encyclopédie Bordas* Volume 7, 1972, p. 179.
88. Roberts, L.G. Machine perception of three-dimensional solids. In *Optical and Electro-optical Information Processing* Massachusetts Institute of Technology Press, 1965.
89. Basseville, M. *The Treatment of Images* IRISA Internal Report 91, 1979.
90. Belan, J.P. *Video Treatment Systems Adapted to Analysis Bars Used in Telescopy* CCETT Internal Report, 1976.
91. Chow, C.K.; Kaneko, I. Boundary detection of radiographic images by a threshold method. *Frontiers of Pattern Recognition* ed. Watanabe, S., 1972.
92. Dessimoz, J.D.; Kunt, M.; Zurcher, J.M. Recognition and handling of overlapping industrial parts. *Proceedings of 9th International Symposium on Industrial Robots* Washington, DC, March, 1979.
93. Freeman, H. On the encoding of arbitrary geometric configurations. *IRE Transactions on Electrical Components* 1961, EC-10, 260–268.
94. Biemond, J.; Gerbrands, J.H. An edge preserving recursive noise smoothing algorithm for image data. *IEEE Transactions on Systems, Man and Cybernetics* 1979, 9(10).
95. Nahi, Y.; Habibi, I. Decision directed recursive image enhancement. *IEEE Transactions on Circuits and Systems* March, 1975.

96. Eberlein, M. An iterative gradient edge detection algorithm. In *Digital Picture Processing* ed. Rosenfield, A., Academic Press, New York, 1976.
97. Modestino, S.; Fries, K. Edge detection in noisy images using recursive digital filtering. *Computer Graphics and Image Processing* 1977, **6**(5).
98. Rosenfield, A.; Thurston, H. Edge and curve detection for visual scene analysis. *IEEE Transactions on Computers* May, 1971.
99. Haralick, H.; Shanmugan, F.; Dinstein, L. Textural features for image classification. *IEEE Transactions on Systems, Man and Cybernetics* 1973, **3**.
100. Ando, S.; Doi, Y. Texture modelling and classification. *Transactions of the Society of Instrument and Control Engineers* February, 1978, pp. 84–89.
101. Hueckel, M. A local visual operator which recognises edges and lines. *Journal of ACM* 1973, **20**(4).
102. Mero, L.; Vassy, Z. A simplified and fast version of the Hueckel operator for finding optimal edges in pictures.
103. Rosenfield, A. The Max Roberts operator is a Hueckel type edge detector. *IEEE Transactions on Pattern Recognition* 1981, **3**(1), 101–103.
104. Griffith, A.K. Edge detection in simple scenes using *a priori* information. *IEEE Transactions on Computers* April, 1973.
105. Giralt, G.; Ghallab, M.; Stuck, F. Object identification with an optimal sequential pattern recognition method. *Proceedings of 9th International Symposium on Industrial Robots* Washington, DC, March, 1979.
106. Brill, E.L. *Character Recognition via Fourier Descriptors* Wescon Paper 25/3, Los Angeles, 1968.
107. Smith, F.; Wright, M. Automatic ship photo interpretation by the method of moments. *IEEE Transactions on Computers* May, 1971.
108. Birk, J.; Kelley, R. *General Methods to Enable Robots with Vision to Acquire, Orient and Transport Workpieces* University of Rhode Island Report, August, 1979.
109. Dudani, S.A. *Moments Methods for the Identification of the Three-dimensional Objects from Optical Sensors* Thesis, Ohio State University, 1971.
110. MacQueen, J.B. Some methods for classification and analysis of multivariate observations. *Proceedings of 5th Berkeley Symposium on Mathematical Statistics and Probability* University of California, 1967.
111. Coiffer, P.; Rives, P. An experiment on vision-movement interaction in automatic handling by robots. *Proceedings of 10th International Symposium on Industrial Robots* Milan, 1980.
112. Coiffet, P.; Rives, P. Robot recognition of the orientation of three-dimensional objects, for automatic gripping tasks. *RAIRO Automatique* (Systems Analysis and Control), February, 1980.
113. Calabi, L.; Hartnett, W.E. Shape recognition, prairie fires, convex deficiencies and skeletons. *American Mathematics Monthly* 1968, **75**, 335–342.
114. Muhlenfeld, E. An optical sensor for locating objects. *Proceedings of 1st CISM-IFTOMH* (International Centre for Mechanical Sciences/International Federation for the Theory of Machines and Mechanisms) Udine, September, 1973.
115. Ball, G.H. *The Application of Integral Geometry to Machine Recognition of Visual Patterns* Wescon Paper 6/3, August, 1962.
116. Chow, C.K. A recognition method using neighbour dependence. *IRE Transactions on Electronic Computers* October, 1962.
117. Aguillar-Martin, J. Iterative classification algorithms in the absence of initial information. *Cybernica* 1972, no. 4.
118. Fix, E.; Hodges, J.L. Discriminatory analysis. Non-parametric discrimination constancy properties. *IEEE Press Machine Recognition of Patterns* 1977.

119. Rosenbaltt, M. Remarks of some non-parametric estimates of a density function. *Annals of Mathematical Statistics* 1956, 27.
120. Parzen, E. On estimation of a probability density function and mode. *Annals of Mathematical Statistics* 1962, 33.
121. Loftsgaarden, D.O.; Quesenberry, C.P. A non-parametric estimate of a multivariate density function. *Annals of Mathematical Statistics* 1965, **36**, 1049–1051.
122. Pot, J.; Rives, P. The problem of vision in industrial robotic applications. *Deuxième Journée de Robotique* Rennes, May, 1981.
123. Pot, J. Thesis, Montpellier, in preparation.
124. Hanson, A.R.; Riseman, E.M. *Computer Vision Systems* Academic Press, New York, 1978.
125. Gleason, G.J.; Agin, G.J. A modular vision system for sensor-controlled manipulation and inspection. *Proceedings of 9th International Symposium on Industrial Robots* Washington, DC, March, 1979, pp. 57–70.
126. Karg, R.; Lanz, O.E. Experiment results with a versatile optoelectronic sensor in industrial applications. *Proceedings of 9th International Symposium on Industrial Robots* Washington, DC, March, 1979, pp. 247–264.
127. Barbera, A.J.; Albus, J.S.; Fitzgerald, M.L. Hierarchical control of robots using microcomputers. *Proceedings of 9th International Symposium on Industrial Robots* Washington, DC, March, 1979, pp. 405–422.
128. Eskenazi, R.; Wilf, J.M. *Low Level Processing for Real-time Image Analysis* NASA, JPL Publication, September, 1979.
129. Roesser, R.P. Two-dimensional microprocessor pipelines for image processing. *IEEE Transactions on Computers* February, 1978.
130. Van Hoorhis, D.C.; Morrin, J.H. in *IEEE Transactions on Computers* February, 1978.
131. Agin, G. An experimental vision system for industrial applications. *Proceedings of 5th International Symposium on Industrial Robots* Chicago, 1975.
132. Agin, G. Servoing with visual feedback. *Proceedings of 7th International Symposium on Industrial Robots* Tokyo, 1977.
133. Chien, R.T.; Jones, V.C. Acquisition of moving objects and hand-eye coordination. *Proceedings of 4th International Joint Conference on Artificial Intelligence* Tbilissi, Georgia, USSR, September, 1975.
134. Hale, J.A.; Saraga, P. Control of a PCB drilling machine by visual feedback. *Proceedings of 4th International Joint Conference on Artificial Intelligence* Tbilissi, Georgia, USSR, September, 1975.
135. Kashioka, S.; Takeda, S.; Shima, Y.; Uno, T.; Hamada, T. An approach to the integrated intelligent robot with multiple sensory feedback. Visual recognition techniques. *Proceedings of 7th International Symposium on Industrial Robots* Tokyo, 1977.
136. Lubinski, K. Microprocessor-based interface converts video signal for objects tracking. *Computer Design* December, 1977.
137. Niemi, A.; Malinen, P.; Koskinen, K. Digitally implemented sensing and control function for a standard industrial robot. *Proceedings of 7th International Symposium on Industrial Robots* Tokyo, 1977.
138. Perkins, W.A. A model based vision system for industrial parts. *IEEE Transactions on Computers* February, 1978.
139. Salmon, M. New part orienting device. *Proceedings of 7th International Symposium on Industrial Robots* Tokyo, 1977.
140. Tani, K.; Abe, M.; Tanie, K.; Ohno, T. High precision manipulator with visual sense. *Proceedings of 7th International Symposium on Industrial Robots* Tokyo, 1977.

141. Tsuji, S.; Nakamura, A. Recognition of an object in a stock of industrial parts. *Proceedings of 4th International Joint Conference on Artificial Intelligence* Tbilissi, Georgia, USSR, September, 1975.
142. Yachida, M.; Tsuji, S. A machine vision system for complex industrial parts with learning capabilities. *Proceedings of 4th International Joint Conference on Artificial Intelligence* Tbilissi, Georgia, USSR, September, 1975.
143. Yachida, M.; Tsuji, S. A versatile machine vision system for complex industrial parts. *IEEE Transactions on Computers* September, 1977.
144. Daly, K. Vision systems suitable for use in robots. In *Assisted System Conception* DEA Report, Montpellier, 1980.
145. Kanade, T. Model representations and control structures in image understanding. *Proceedings of 5th International Joint Conference on Artificial Intelligence* September, 1977, pp. 1074–1082.
146. Huffman, D.A. Impossible objects as nonsense sentences. *Machine Intelligence* 1971, **6**, 295–323.
147. Waltz, D.L. Understanding line drawings of scenes with shadows. In *The Psychology of Computer Vision* McGraw-Hill, New York, 1975.
148. Huffman, D.A. A duality concept for the analysis of polyhedral scenes. *Machine Intelligence* 1977, **8**, 475–492.
149. Falk, G. Interpretation of imperfect line data as a three-dimensional scene. *Artificial Intelligence* 1972, 3-2.
150. Turner, K.J. *Computer Perception of Curved Objects Using a TV Camera* Thesis, School of Artificial Intelligence, Edinburgh, 1974.
151. Rhodes, M.L.; Glenn, W.V.; Klinger, A. Three-dimensional isolation structure using parallel image planes. *Proceedings of 4th IJCPR* 1979, 584–591.
152. Shirai, Y. A computer system for recognition of real world objects. *Report no. 13*, Electromechanical Laboratory, Tokyo, 1977.
153. Gennery, D. A stereo vision system for autonomous vehicles. *Proceedings of 5th International Joint Conference on Artificial Intelligence* September, 1977, pp. 576–582.
154. Yakimovski, Y.; Cunningham, R. A system for extracting 3D measurements from a stereo piece of TV cameras. *Computer Graphics and Image Processing* Volume 7, April, 1978.
155. Driscoll, L.C.; Levine, J. A contour recognition method for identifying parts/part orientation in automatic assembly. *Proceedings of 5th International Joint Conference on Artificial Intelligence* 1975, pp. 371–386.
156. Kelley, R.; Birk, J.; Wilson, L. An algorithm to visually acquire workpieces. *Proceedings of 7th International Symposium on Industrial Robots* Tokyo, 1977.
157. Romeder, J.M. *Discriminatory Method and Programme Analysis* Dunod, 1973.

Index